BRAIN FICTION

Philosophical Psychopathology: Disorders in Mind
Owen Flanagan and George Graham, editors

Divided Minds and Successive Selves: Ethical Issues in Disorders of Identity and Personality
Jennifer Radden

The Myth of Pain
Valerie Hardcastle

Imagination and the Meaningful Brain
Arnold Modell

Brain Fiction: Self-Deception and the Riddle of Confabulation
William Hirstein

BRAIN FICTION
Self-Deception and the Riddle of Confabulation

William Hirstein

A Bradford Book
The MIT Press
Cambridge, Massachusetts
London, England

First MIT Press paperback edition, 2006

MIT Press books may be purchased at special quantity discounts for business or sales promotional use. For information, please email special_sales@mitpress.mit. edu or write to Special Sales Department, The MIT Press, 55 Hayward Street, Cambridge, MA 02142.

This book was set in Stone serif and Stone sans on 3B2 by Asco Typesetters, Hong Kong and was printed and bound in the United States of America.

Library of Congress Cataloging-in-Publication Data

Hirstein, William.
Brain fiction : self-deception and the riddle of confabulation / William Hirstein.
   p.   cm. — (Philosophical psychopathology. Disorders in mind)
"A Bradford book."
Includes bibliographical references and index.
ISBN-13: 978-0-262-08338-6 (hc. : alk. paper) — 978-0-262-58271-1 (pb. : alk. paper)
ISBN-10: 0-262-08338-8 (hc. : alk. paper) — 0-262-58271-6 (pb. : alk. paper)
1. Deception. 2. Self-deception. 3. Mythomania. I. Title. II. Series.
RC569.5.D44H577   2005
615.85′84—dc21   2004050498

10 9 8 7 6 5 4 3 2

Contents

Series Foreword   vii
Acknowledgments   ix

1   What Is Confabulation?   1
1.1   Introduction   1
1.2   Confabulation Syndromes   7
1.3   Features of Confabulation   15
1.4   Three Concepts of Confabulation   19
1.5   Mirror-Image Syndromes   21
1.6   Conclusion: Setting the Problem of Confabulation   22

2   Philosophy and Neuroscience   25
2.1   The Growth of Neuroscience   25
2.2   Principles of Brain Structure and Function   31
2.3   The Limbic and Autonomic Systems   36
2.4   Philosophy's Role   37
2.5   Approach of This Book   40

3   Confabulation and Memory   43
3.1   Fictional Autobiographies   43
3.2   The Brain's Memory Systems   46
3.3   Korsakoff's Syndrome   49
3.4   Aneurysms of the Anterior Communicating Artery   55
3.5   Frontal Theories of Confabulation   60
3.6   Separating Amnesia and Confabulation   65
3.7   False Memories   66
3.8   Conclusion   67

4   Liars, Sociopaths, and Confabulators   71
4.1   Unhappy Family: The Orbitofrontal Syndromes   71
4.2   Symptoms of Orbitofrontal Damage   74
4.3   Anatomy and Physiology of the Orbitofrontal Cortex   82
4.4   Sociopathy   90
4.5   Lying and the Skin-Conductance Response   93
4.6   Obsessive-Compulsive Disorder as a Mirror-Image Syndrome   97
4.7   Conclusion   99

5   Mind Reading and Misidentification   101
5.1   Knowledge of Others' Minds   101
5.2   Mind-Reading Systems   104
5.3   Misidentification Syndromes   114
5.4   A Mind-Reading Theory of Misidentification   122
5.5   Conclusion   133

6   Unawareness and Denial of Illness   135
6.1   Denial   135
6.2   Theories of Anosognosia   140
6.3   The Neuroscience of Denial   143
6.4   Denial of Blindness   145
6.5   Anosognosia and the Other Confabulation Syndromes   147
6.6   Conclusion   151

7   The Two Brains   153
7.1   Confabulations by Split-Brain Patients   153
7.2   Hemispheric Differences   155
7.3   Anatomy of the Cerebral Commissures   158
7.4   Lateral Theories of Confabulation   160
7.5   Evaluating the Lateral Theories   166
7.6   Other Confabulations about Mental States and Intentions   170
7.7   Conclusion   176

8   Confabulation and Knowledge   177
8.1   Confabulation as an Epistemic Phenomenon   177
8.2   The Neuroscience of Confabulation   179
8.3   Creation and Checking of Mental Representations   181
8.4   Defining Confabulation   187
8.5   Other Candidate Criteria and Conceptions   198
8.6   Epistemic Features of Confabulation   203
8.7   Knowing That We Do Not Know   209
8.8   Conclusion   211

9   Self-Deception   213
9.1   Confabulation: Clues to Self-Deception   213
9.2   Deception and Lying   218
9.3   What Is Self-Deception?   221
9.4   The Maintenance of Self-Deceptive Beliefs   226
9.5   Questions about Self-Deception   228
9.6   Self-Deception and Mind Reading   233
9.7   The Neuroscience of Self-Deception   234
9.8   Conclusion   236

10   Epilogue: Our Nature   239
10.1 The Meaning of Confabulation   239
10.2 Further Questions   241

References   245
Name Index   277
Subject Index   283

Series Foreword

The aim of this series is both interdisciplinary and uncharted: to offer a philosophical examination of mental disorders, an area of intense and fascinating activity in recent years. The perspective of philosophy provides a richly synoptic vision of the forms, limits, and lessons of mental disorders, as well as their study and treatment. Potential topics include, but are not limited to, the following:

- How to explain mental disorders
- Dissociative personality and volitional disorders and what they tell us about rational and moral agency
- Disorders of consciousness and what they tell us about selfhood and self-representation
- Lessons of cognitive neuropsychology for the nature and function of mind
- Whether disorders are "rational strategies" for coping with trauma or stress
- Relationships between dream states and psychosis
- Neural-network models of pathology and their implications for the debate over the functional integration of mind and brain
- Culture-specific and gender-linked forms of psychopathology and their lessons for both the taxonomy of mental disorders and the scientific status of the study of mental illness
- Logical and epistemological relations between theories of mental disorders and forms of therapy
- Conceptual and methodological foundations of psychopharmacology
- Ethical and political issues in definition and treatment of mental disorders

The editors welcome proposals and submissions from philosophers, cognitive scientists, psychiatric researchers, physicians, social scientists, and others committed to a philosophical approach to psychopathology.

Owen Flanagan
George Graham

Acknowledgments

My first exposure to a confabulating patient occurred when I went with V. S. Ramachandran, who was then my postdoctoral supervisor, to a rehabilitation center near the University of California, San Diego. We visited an older man who had been an English professor and who had recently had a serious stroke. We knew that the stroke had paralyzed his left arm, but when we asked him about the strength in his arms, he calmly replied that they were fine. His complete confidence in his false answers to this and other questions was striking, given his basic mental soundness and reasonableness in other areas. This made me wonder about what exactly was wrong with him, and what confabulation might indicate about the structure of the human mind.

Once I learned about confabulation from observing patients with neurological disorders, I began to see hints of it in normal people. It started to become apparent that what we were seeing in the patients was an extreme version of some basic feature of the human mind, having to do with the way we form beliefs and report them to others. If this is true, there should be much to learn from a study of confabulation, or at least that is what I argue here.

I would like to thank the following people who influenced my thinking and educated me on the different topics discussed here: Eric Altschuler, Sandra Bluhm, William Bossart, William Brenner, Melinda Campbell, Patricia Churchland, Paul Churchland, Russell Goodman, George Graham, Portia Iversen, Brian Keeley, G. J. Mattey, Sandra de Mello, Thomas Natsoulas, Helga Noice, V. S. Ramachandran, G. Fred Schueler, Nadia Sahely, John R. Searle, and Richard Wollheim. Thanks to Joel Huotari, David Metcalfe, and Darryl Thomas for assisting with the bibliography. Thanks also to Kathy Willis and the staff at the Buehler Library for their relentless efforts in tracking down books and articles. The original figures and figure adaptations are by Katie Reinecke.

BRAIN FICTION

1 What Is Confabulation?

They talk freely in the intervals of mitigation, but of things that do not exist; they describe the presence of their friends, as if they saw realities, and reason tolerably clearly upon false premises.
—John Coakley Lettsom (1787)

1.1 Introduction

A neurologist enters a hospital room and approaches an older man sitting up in bed. The neurologist greets him, examines his chart, and after a brief chat in which the man reports feeling fine, asks him what he did over the weekend. The man offers in response a long, coherent description of his going to a professional conference in New York City and planning a project with a large research team, all of which the doctor writes down. The only problem with this narration is that the man has been in the hospital the entire weekend, in fact for the past three months. What is curious is that the man is of sound mind, yet genuinely believes what he is saying. When the doctor informs him that he is mistaken, he replies, "I will have to check with my wife about that," then seems to lose interest in the conversation. The man isn't "crazy" or schizophrenic; he is quite coherent and can answer all sorts of questions about who his children are, who the current president is, and so on. He is *confabulatory*, owing in this case to the fact that he has Korsakoff's syndrome, a disorder that affects his memory, producing a dense amnesia for recent events. But unlike other patients with memory dysfunction, who freely admit their memories are poor, a patient with Korsakoff's syndrome will confidently report as memories events that either did not happen (or at least did not involve him) or that happened to him, but much earlier in life. This man's act of describing the conference in New York City is known as a *confabulation*.

The neurologist moves down the hall to a room in which an older woman patient is in bed talking to her daughter. When the daughter sees him, she asks to speak with him outside in the hallway. "She won't admit she's paralyzed, Doctor. What's *wrong* with her?" The woman has a condition familiar to anyone who has worked with people after a stroke, known as denial of paralysis (hemiplegia). The more general name for this condition is *anosognosia*, which means lack of knowledge about illness. It can come about when a stroke damages a certain part of the right hemisphere just behind and above the right ear, causing paralysis or great weakness on the left side of the body. The denial tends to occur right after the patient recovers consciousness, and tends to last only a few days.

The doctor walks back in the room, approaches the woman, and greets her. When he asks her how she is, she reports feeling fine. "Are both your hands equally strong, Mrs. Esposito?" he asks. "Yes they're fine," she replies. "Can you touch my nose with your right hand?" he asks. She reaches up, a bit unsteadily, but succeeds in touching the doctor's nose. "Would you touch my nose with your left hand?" he then asks. Mrs. Esposito pauses a moment, rubs her left shoulder and replies, "Oh, I've got severe arthritis in my shoulder. You know that Doctor; it hurts." Again, she is not lying or pretending, she genuinely believes that she can move her arm. She also believes her confabulation about arthritis.

Perhaps what is most troubling about witnessing such confabulations is the rock-jawed certainty with which they are offered up. The patients give none of the outward appearances of lying, and indeed most writers on the subject do not consider confabulation to be lying, because it lacks at least two crucial components: the intent to deceive, and knowledge contrary to what is claimed. The claims about arthritis or the conference in New York City were not intentionally concocted with the motive of deception in mind; the patient is reporting what seems true to him or her.

Why then does confabulation happen? Confabulation seems to involve two sorts of errors. First, a false response is created. Second, having thought of or spoken the false response, the patient fails to examine it and recognize its falsity. A normal person, we want to say, would notice the falsity or absurdity of such claims. The patient should have either not created the false response or, having created it, should have censored or corrected it. We do this sort of censoring in our normal lives. If I ask you whether you have ever been to Siberia, for instance, an image might appear in your mind of you wearing a thick fur coat and hat braving a snowy storm, but you know that this is fantasy, not reality. In very general terms, the confabulating patient lacks the ability to assess his or her situation, and to either answer correctly, or respond that he or she does not know. Apparently, admitting ignorance in response to a question, rather than being an indication of glibness and a low level of function, is a high-level cognitive ability, one that confabulators have lost. "I don't know," can be an intelligent answer to a question, or at least an answer indicative of good cognitive health.

Confabulation was initially considered solely a disorder of memory in which a patient gives false or contrived answers to questions about his orher past, but believes those answers to be true, as in the case of the man with Korsakoff's syndrome. "Confabulation" as a technical term was applied first to Korsakoff's patients by the German psychiatrist Karl Bonhoeffer in the early 1900s (Berrios 1998). In this narrower conception, confabulation occurs when patients produce stories that fill in gaps in their memories. The American Psychiatric Association's official *Diagnostic and Statistical Manual of Mental Disorders* (known as DSM IV), for instance, defines confabulation as "the recitation of imaginary events to fill in gaps in memory"

(1994, 433). However, confabulation also appears in a wide variety of other syndromes, many of which involve no obvious memory problems, including as we saw, anosognosia for hemiplegia (denial of paralysis), but also split-brain syndrome, Anton's syndrome (denial of blindness), Capgras' syndrome (the illusion that an impostor has replaced a person close to the patient), and schizophrenia.

The apparent diversity of confabulation syndromes invites a search for something they have in common. If a single brain region is damaged in all of these patients, and we know something about the function of that part of the brain, perhaps this knowledge can allow us to begin to unravel the mystery of why people confabulate and what it tells us about brain function. Unfortunately, it is not that simple because the sites of damage in confabulating patients seem to be widely scattered throughout the brain. Nevertheless, theories of the sites of lesions in confabulation have had two rough areas of focus: the frontal lobes and the corpus callosum—the large bundles of nerve fibers that interconnect the left and right hemispheres (Pandya and Seltzer 1986). Most recently, accounts of the locus of lesion in confabulation have tended to focus on the frontal lobes (Stuss et al. 1978; Benson et al. 1996; Burgess and Shallice 1996; Johnson et al. 1997). These memory-based frontal theories are examined in chapter 3. It has long been suspected, however, that the neural locus of confabulation has an important lateral component, owing to the presence of confabulation in split-brain patients (see chapter 7). The fact that denial of paralysis happens overwhelmingly with right hemisphere strokes also seems to indicate a lateral component in confabulation (see chapter 6).

Confabulation has also been reported in young children reporting their memories, in subjects of hypnosis, and in normal people in certain experimental settings. When normal people are asked about certain choices they made, they can produce something that sounds rather like a confabulation. Nisbett and Wilson (1977) set up a table in a department store with pairs of nylon stockings and asked shoppers to select the pair they preferred. Unbeknown to the shoppers, the pairs were identical. People tended to choose the rightmost pair for reasons that are not clear, but when asked the reason for their choice, the shoppers commented on the color and texture of the nylons. When they were told that the nylons were identical, and about the position effects, the shoppers nevertheless tended to resist this explanation and stand by their initial reasons. As with patients with neurological disease, the question that is raised by such behavior is, why didn't the shoppers reply that they didn't know why they preferred that pair of nylons?

Rather than being merely an odd neurological phenomenon, the existence of confabulation may be telling us something important about the human mind and about human nature. The creative ability to construct plausible-sounding responses and some ability to verify those responses seem to be separate in the human brain. Confabulatory patients retain the first ability, but brain damage has compromised the second. One of the characters involved in an inner dialogue has fallen silent as the other rambles on unchecked, it appears. Once one forms a concept of confabulation from seeing it in the clinic or reading about it in the neuropsychological literature, one starts to see mild versions of it in normal people. We are all familiar with people who seem to be unable to say the words, "I don't know," and will quickly produce some sort of plausible-sounding response to whatever they are asked. A friend once described a mutual acquaintance as "a know-it-all who doesn't know anything." Such people have a sort of mildly confabulatory personality, one might say. One soon learns to verify any information they offer, especially if it involves something important.

One way to connect these normal cases of confabulatory people with the clinical examples is the idea that the normal people may be the same sort of people who exhibit clinical-level confabulation after brain injury. In a study of the premorbid personality of his anosognosic patients, Weinstein reported that "relatives saw them as stubborn, with an emphasis on being right" (1996, 345). Confabulation in the clinic might be produced by the suddenness of the injury, but on the normal spectrum, there may be all sorts of mild degrees of it among us. Those with clinical confabulation slowly learn to check, doubt, and finally inhibit their confabulations; similarly, normal people may become mildly confabulatory for a period as they age, then learn how to correct for it.

Confabulation involves absence of doubt about something one should doubt: one's memory, one's ability to move an arm, one's ability to see, and so on. It is a sort of pathological certainty about ill-grounded thoughts and utterances. The phenomenon contains important clues about how humans assess their thoughts and attach either doubt or certainty to them. Our expressions of doubt or certainty to others affect how they hear what we say. In the normal social milieu, we like people to be certain in their communications, and strong social forces militate against doubting or pleading ignorance in many situations. A cautious weather forecaster, for example, who when asked about tomorrow's weather always replies, "I don't know," or "Can't be sure," will soon be unemployed. Excessive caution is especially counterproductive when it occurs in someone in a position of power or responsibility. Imagine a general who was never sure what to do, and as a consequence never did anything, never commanded his soldiers because he always felt that he didn't know for certain what the right strategy was. Armies (and life forms) that do not move forward aggressively are soon overtaken by their enemies or competitors. Those under the authority of the leader can find admissions of ignorance especially troubling, even frightening. Imagine the president being asked what he plans to do about a current oil crisis, for instance, and his answering, "I don't know what to do about it." The stock market would plummet. At least in some contexts, then, an answer that is possibly (or even probably) wrong is better than none at all.

There is also a clear connection here to the human gift for storytelling. Many confabulations are plausible little stories, about what one did over the weekend, or why one can't move one's arm. We all have little stories we tell ourselves and others, especially when we are asked why we did something. Lovers, for instance, are notorious for asking, "Why do you love me?" Often we are not really sure—we simply are drawn to the person; so much of what is important in relationships happens below the level of explicit awareness. However, we usually offer up some account: "I like your eyes," or "I like your enthusiasm." We also each have a sort of personal story that we tell to ourselves and others—about how interesting, successful, ethical, honest, etc., we are. Are these phenomena at the normal end of a continuum, with confabulating people with neurological disorders at the other? Philosopher Daniel Dennett argued that one sort of meaning we can give to the overworked term "the self" is that the self is the subject of a story we create and tell to others about who we are: "Our fundamental tactic of self-protection, self-control, and self-definition is not spinning webs, but telling stories, and more particularly concocting and controlling the story we tell others—and ourselves—about who we are" (Dennett 1991, 418).

These stories have a unifying function according to Dennett: "These strings or streams of narrative issue forth *as if* from a single source—not just in the obvious physical sense of flowing from just one mouth, or one pencil or pen, but in a more subtle sense: their effect on any audience is to encourage them to (try to) posit a unified agent whose words they are, about whom they are: in short, to posit a center of narrative gravity" (1991, 418). Perhaps confabulation arises in part from this natural inclination toward telling stories about ourselves. Confabulation may also share with Dennett's storytelling an unintentional quality: "And just as spiders don't have to think, consciously and deliberately, about how to spin their webs, and just as beavers, unlike professional human engineers, do not consciously and deliberately plan the structures they build, we (unlike *professional* human storytellers) do not consciously and deliberately figure out what narratives to tell, and how to tell them" (1991, 418). In this Dennettian conception, confabulation is also a social phenomenon, but one that is more directly in the service of the individual than those mentioned earlier about the need for leaders to be confident. We will examine this view of confabulation in greater detail in chapter 7.

In an alternative conception, confabulation reflects the need that people have to assess their current situations and act on them quickly, without pausing to consider all possibilities (Ramachandran and Blakeslee 1998). One of the brain's primary functions is to make sense of the world. Things and people must be categorized as this or that, as Tom or Mary, and decisions must be made about what to do with these things and people. Making categorizations always involves a risk of error, but doubt is a cognitive luxury and occurs only in highly developed nervous systems. One cannot focus too much on the details of what exactly the situation is, otherwise the most important issues are lost. Assuming it *is* a tiger rustling about in the bushes, what should I do about it? The brain has a proclivity to smooth over the rough edges and ignore certain details so as not to lose the big picture. Perhaps confabulation is a result of this sort of engineering strategy, but on a larger scale.

The brain likes to fill in gaps in the visual field, for instance. Each of your eyes has a blind spot, above and to the outside of that eye's focal center, where there are no rods or cones to receive light because the optic nerve exits the eyeball there. The reason you don't perceive a black spot in this portion of your visual field is that parts of the visual cortex fill in the blind spot, based on what is being perceived around its periphery (Ramachandran and Churchland 1994). For instance, if a solid patch of red is perceived around the periphery, these processes will fill in the blind spot with red, even though there might actually be a green spot there. Presumably the brain engages in these sorts of "coherencing" processes to facilitate its primary work: selection of actions from perceptions. Is the filling in of gaps in memory similar to filling in the blind spot, as Sully speculated long ago? "Just as the eye sees no gap in its field of vision corresponding to the 'blind spot' of the retina, but carries its impression over this area, so memory sees no lacuna in the past, but carries its image of conscious life over each of the forgotten spaces" (1881, 282). Does confabulation then belong with filling in as one of these "coherencing" or smoothing processes?

Aside from the questions about the causes of confabulation and its implications for our understanding of the mind, there are also practical concerns that motivate our looking into the problem. In patients with neurological disease, confabulation is a barrier to the rehabilitation processes necessary for recovery, since they will not try to improve their memories, or their range of motion, until they acknowledge that a problem exists. The rest of us also need to know if we are prone to a problem that causes us to chronically misrepresent the world, and misunderstand and misrepresent to others how much we really know about ourselves or why we do things.

The Etymology of "Confabulation"

According to the *Oxford English Dictionary* (OED 1971), "confabulation" is descended from the Latin term *confabulari*, which is constructed from the roots *con*, meaning together, and *fabulari*, which means to talk or chat, so that the original meaning of "to confabulate," is to talk familiarly with someone. Edgeworth wrote in 1801, for example, "His lordship was engaged in confabulation with his groom." Confabulations are "talkings," as Kempis translated in 1450: "consolacions are not as mannes talkinges or confabulacions." The OED also notes, however, that *fabula* means tale, and evolved into the English word "fable." So that "to confabulate with" another person is "to fable with" that person, as it were.

At the turn of the century neurologists began applying a different sense of "confabulation" to some of their patients, beginning with those exhibiting what later came to be called Korsakoff's amnesia, as in the example at the beginning of this chapter (Bonhoeffer 1904; Pick 1905; Wernicke 1906). The precise definition of the neurologist's use of the word has been a subject of debate since it was introduced, however, owing to a lack of consensus as to what exactly the different essential features of confabulation are (Berlyne 1972; Whitlock 1981; Berrios 1998).

1.2   Confabulation Syndromes

It is widely accepted that confabulation comes in two forms, a milder version that includes the above examples, and a more severe and rare version in which the patient invents fantastic or absurd stories. The milder version was initially referred to as "momentary confabulation" by Bonhoeffer in his writings on patients with Korsakoff's syndrome (1901, 1904). He called these "confabulations of embarrassment," and speculated that they are created to fill in gaps in memory. Alternatively, what he referred to as "fantastic confabulation" overlaps heavily with things people who are under the influence of delusions say. Fantastic confabulations tend to have strange, "florid" (Kopelman 1991), or "extraordinary" (Stuss et al. 1978) content. Kopelman maintained, however, that the momentary-fantastic distinction "confounds a number of factors, which are not necessarily correlated, as the distinguishing features of the two types of confabulation; and it is wiser, perhaps, to focus attention upon one central feature by referring to 'provoked' and 'spontaneous' confabulation" (1987, 1482).

Provoked confabulation is produced in response to a question, whereas spontaneous confabulators produce their confabulations without being asked. Kopelman's provoked–spontaneous dichotomy has gained some acceptance, and a body of recent research makes use of those concepts, for instance, in discussing whether they involve separate sites of damage (see Fischer et al. 1995; Schnider et al. 1996). An additional related question addresses whether the two types are discrete or merely represent gradations of severity. DeLuca and Cicerone (1991) resisted Berlyne's claim that "fantastic confabulation seems to be a distinct entity having nothing in common with momentary confabulation" (Berlyne 1972, 33), holding that the two types exist on a continuum. Consistent with this, there are reports of patients in whom spontaneous confabulation became provoked confabulation as the patients improved (Kapur and Coughlan 1980).

We will focus primarily on the provoked or milder form of confabulation, for a number of reasons. What is fascinating about provoked confabulation is that it occurs in people who are fully in possession of most of their cognitive faculties, and able to respond correctly to all sorts of requests and questions. Spontaneous confabulations are irrational stories that presumably result from delusions, and are seen primarily in schizophrenics. The value of provoked confabulation is that it shows the cognitive apparatus malfunctioning in an otherwise sound mind. It also promises to reveal valuable insights about the functioning of the normal cognitive system. The cognitive system of the schizophrenic confabulator is so severely broken that it is much more difficult to glean insights from it.

Before we take a closer look at the definition of confabulation we need to see some of the raw data. What follows is a survey of the different neurological syndromes known to produce confabulation, with several examples both from the clinic and from everyday settings.

Confabulation syndromes seen in the clinic include:
- Korsakoff's syndrome
- Aneurysm of the anterior communicating artery (ACoA)
- Split-brain syndrome
- Anosognosia for hemiplegia
- Anton's syndrome
- Capgras' syndrome
- Alzheimer's disease
- Schizophrenia

Confabulation has also been reported in normal people in certain circumstances, including young children, subjects of hypnosis, people asked to justify opinions, and people reporting mental states.

Confabulation in Clinical Settings
Korsakoff's syndrome is a form of amnesia, most often caused by a lifetime of heavy drinking (Korsakoff 1889; Kopelman 1987). The memory deficit typically affects autobiographical memory (our memory of what happened to us) most severely, but it can also affect semantic memory (our impersonal knowledge of concepts and facts). The most frequent sites of damage are the mamillary bodies and the dorsomedial nuclei of the thalamus. Korsakoff's amnesia is severe enough that the patient will typically have no

memory at all of the events of the preceding day. However, when asked what he or she did yesterday, the Korsakoff's patient will often produce a detailed description of plausible (or not so plausible)-sounding events, all of it either entirely made up on the spot, or traceable to some actual but much older memory.

Berlyne gave the following example of confabulation from a patient with Korsakoff's syndrome: "L.M. was a 46-year old chronic alcoholic. He was admitted having been found at home in a very neglected state. He was euphoric, and showed devastation of recent memory. He said that he had been brought to the hospital by two Sick Berth Attendants and a Petty officer and that he was serving at Gosport on M.T.B.'s. (He had in fact been a Chief Petty Officer on motor torpedo boats during the war.) He said that the war was on and the invasion of Europe was imminent, yet he could recollect both VE and VJ day. He gave the date correctly; when told this would mean that the war had been going on for 20 years he was unperturbed" (Berlyne 1972, 32). Confabulatory patients often contradict themselves and, as with this man, show no interest in reconciling the contradictory claims.

Aneurysms of the anterior communicating artery, which distributes blood to a number of structures in the front of the brain, frequently produce confabulation (DeLuca and Diamond 1993; Fischer et al. 1995). They also produce a dense amnesia similar to that in Korsakoff's syndrome, leading some authors to speculate that confabulation is a deficit in monitoring information about "source memory"—the ability to place where in time and space a remembered event occurred—since the two syndromes share this feature.

The following is a description by Johnson and co-workers of a confabulating patient with an anterior communicating artery aneurysm: "G.S. had a number of erroneous ideas about his personal life to which he clung despite attempts to dissuade him. His fabrications were generally plausible and many involved autobiographical events that were embellished. A particularly salient confabulation was G.S.'s account of the origin of his medical condition; he believed that he had fallen and hit his head while standing outside talking to a friend, when in fact his aneurysm ruptured following an argument with his daughter" (Johnson et al. 1997, 192).

Aside from amnesia and confabulation, the third feature of ACoA syndrome is described as "personality change" (DeLuca 1993), and as with anosognosia, ACoA syndrome is accompanied by unawareness, in this case of the memory problem. Chapter 3 contains a detailed examination of this syndrome and its role as the primary piece of evidence for frontal theories of confabulation. It also contains a detailed inquiry into the locus of the lesion in Korsakoff's syndrome and any overlap it might have with lesion sites in ACoA syndrome.

Split-brain patients are epileptics who have had their corpus callosum surgically removed to lessen the severity and frequency of their seizures. In early testing of these patients in the 1960s, techniques were developed to send visual input to only one hemisphere or the other. Only the left hemisphere was thought able to give verbal responses, but it was found that the right hemisphere could understand pictorial and simple linguistic input, and could respond by pointing to pictures with the left hand [each hemisphere has control over the arm on the opposite side (Gazzaniga 1995a)]. When patients were asked about the activities of the left hand, though, the left hemisphere would answer as if *it* had been controlling the left hand, whereas, as a result of the operation, the left hemisphere had no idea why the left hand was doing what it was.

Several times during testing of the split-brain patients the left hemisphere produced reasonable-sounding but completely false explanations of the left hand's pointing behavior. In one study, a picture of a snow scene was lateralized to the right hemisphere of patient P.S., while a picture of a chicken claw was displayed to his left hemisphere. Then an array of possible matching pictures was shown to each hemisphere. P.S. responded correctly by pointing at a snow shovel with his left hand and at a chicken with his right hand. But when he was asked why he had chosen these items, he responded, "Oh, that's simple. The chicken claw goes with the chicken, and you need a shovel to clean out the chicken shed" (Gazzaniga 1995a, 225). In another study, a picture of a naked woman was shown only to the right hemisphere, using a tachistoscope (Gazzaniga and LeDoux 1978, 154). When the patient was asked why she was laughing, the left hemisphere responded, "That's a funny machine."

Anosognosia is exhibited by people with many types of neurological disorders, but it occurs most frequently after stroke damage to the inferior parietal cortex of the right hemisphere (Bisiach and Geminiani 1991; Heilman et al. 1991). Damage here can produce paralysis or great weakness of the left arm or of the entire left side of the body. This paralysis can be accompanied by neglect, a condition in which the patient ignores the left side of the body and its surrounding space. A patient with neglect typically will not eat food on the left side of her plate or wash the left side of her body, and will not notice people standing quietly on her left. Some patients with this left-side paralysis and neglect also exhibit anosognosia for several days after a stroke. Approached on her right side as she lies in bed and asked whether she can use her left arm, such a patient will answer matter-of-factly that she can. When the neurologist tells the patient to touch his nose with her left arm, the patient will try in vain to reach it, or will occasionally reach out with her right arm instead. But often, she will produce a confabulation, saying something such as, "I have never been very ambi-dextrous," or "These medical students have been probing me all day and I'm sick of it. I don't want to use my left arm" (Ramachandran 1996b, 125).

When asked whether she reached successfully, the patient who tried to reach will often say that she did, and a large percentage of these patients will claim that they saw their hands touch the doctor's nose.

Neglect and denial can also be accompanied by *asomatognosia*, in which the patient denies that the paralyzed or greatly weakened arm even belongs to him or her. Sandifer (1946, 122–123) reported the following dialogue between an examining physician and an anosognosic patient:

Examiner (holding up patient's left hand): "Is this your hand?"
Patient: "Not mine, doctor."
Examiner: "Whose hand is it then?"
Patient: "I suppose it's yours, doctor."
Examiner: "No, it's not; look at it carefully."
Patient: "It is not mine, doctor."
Examiner: "Yes it is, look at that ring; whose is it?"
Patient: "That's my ring; you've got my ring, doctor."
Examiner: "Look at it—it is your hand."
Patient: "Oh, no, doctor."
Examiner: "Where is your left hand then?"
Patient: "Somewhere here, I think." (Making groping movements toward her left shoulder.)

The patients give no sign that they are aware of what they are doing; apparently they are not lying, and genuinely believe their confabulations. They do not give any outward signs of lying, and their demeanor while confabulating was described by Kraepelin (1910) as "rocklike certitude." In one experiment that affirms the sincerity of confabulators, anosognosics with left-side paralysis were given the choice of performing a two-handed task (tying a shoe) for a reward of $10, or a one-handed task (screwing a light bulb into a socket) for $5. The patients uniformly selected, then failed at the two-handed task. In contrast, those who had left-side paralysis caused by a right hemisphere stroke but no anosognosia systematically chose the one-handed task (Ramachandran 1995). In the clinic, anosognosics are often observed trying to use their paralyzed left arms or legs.

The fact that the overwhelming majority of neglect patients ignore the left side of personal space as a result of right hemisphere damage seems to lend support to the claim that the causes of confabulation have an important lateral component. Confabulating neglect patients share with split-brain patients the feature of having a left hemisphere cut off from the information it needs to answer the doctor's questions. Chapter 6 examines these connections after giving a thorough description of what is known about this type of anosognosia.

Another type of anosognosic patient denies that he or she is blind. Known as Anton's syndrome, this is a rare condition that is typically traceable to bilateral damage to the occipital lobes, areas at the back of the brain specialized for visual processing (Anton 1899; Swartz and Brust 1984), coupled with frontal damage. The occipital damage can cause a condition known as cortical blindness—blindness that is due to cortical damage rather than damage to the eyes or optic nerves. A small percentage of patients with cortical blindness also exhibit Anton's syndrome; they tend to have either diffuse damage caused by dementia, or circumscribed frontal lobe lesions (McDaniel and McDaniel 1991). Those with Anton's syndrome can be quite rational in general, until one asks them to describe what they see. They typically produce a description that is logical or plausible, but false. For instance, if asked to describe what their doctor is wearing, they will provide a full description of a generic doctor. When confronted with the falsity of the description, the patients make excuses similar to those of patients who deny paralysis. Benson's patient with Anton's syndrome "adamantly denied any visual problems, often complaining that the light was poor in the room, that he did not have his best pair of glasses with him, or that it was nighttime" (1994, 87).

Capgras' syndrome is a rare condition in which a patient claims that people close to him or her, typically parents, spouses, or children, have been replaced by impostors (Capgras and Reboul-Lachaux 1923). The locus of the lesion is unknown, but a consensus is building that it involves a temporal lobe lesion in concert with a frontal lesion (e.g., Signer 1994). The most popular current hypothesis about why the impostor delusion is formed is that the patient is missing a feeling of emotional arousal that the sight of a significant person normally produces, and that the impostor claim is a confabulation created to explain why the person feels different to the patient. V. S. Ramachandran and I asked a patient with Capgras' syndrome point-blank why someone would pretend to be his father. His reply was: "That is what is so strange, Doctor—why should anyone want to pretend to be my father? Maybe my father employed him to take care of me—paid him some money so he could pay my bills" (Hirstein and Ramachandran 1997, 438). This patient, D.S., also had the same illusion when presented with photographs of himself, as shown by the following interchange (1997):

Examiner (Pointing to photograph of D.S. taken two years ago when he had a moustache): "Whose picture is this?"
Patient: "That is another D.S. who looks identical to me but he isn't me—he has a moustache."

Capgras' syndrome is different from many other confabulation syndromes in that the confabulation posits something different from the status quo, compared with the denials of anosognosics, which seem to be designed to affirm the patient's old body image. The behavior of Capgras' patients tends to be consistent with their impostor claims; there are reports of patients killing the "impostors," for instance.

Alzheimer's disease involves the loss of neurons in several different brain areas, and manifests itself in a progressive loss of abilities in several different cognitive tasks (Wells and Whitehouse 1996). It is the most common type of what neurologists call dementias. Confabulations produced by these patients tend to be similar to those produced by people with Korsakoff's syndrome (Kopelman 1987, 1991). The conventional wisdom is that confabulation occurs in Alzheimer's disease when the diffuse cortical atrophy reaches a point at which the frontal lesions necessary to produce confabulation occur (see Kern et al. 1992), perhaps overlapping with the lesions found in Korsakoff's or in aneurysms of the anterior communicating artery.

Schizophrenia is a serious psychological disorder that involves large-scale disruptions in perception, emotions, thinking, and behavior (Jeste et al. 1996). Unlike some of the other syndromes we have discussed, there is no widely agreed-upon theory as to the locus of damage in the schizophrenic brain. Nathaniel-James and Frith (1996) first broached the idea that confabulation is present in schizophrenics (see also Kramer et al. 1998). They read narratives adapted from Aesop's fables to schizophrenic patients, then immediately afterward asked the subjects to recall as much of the story as they could. When the subjects inserted features that were not present in the story, these were counted as confabulations. Nathaniel-James and Frith suggested that schizophrenics share with frontal damage patients the inability to suppress inappropriate responses, something that links confabulation with the phenomenon of disinhibition (more on this in chapter 4).

Confabulation in Normal People

Young children sometimes confabulate when asked to recall events. Ackil and Zaragoza (1998) showed first-graders a segment of a film depicting a boy and his experiences at summer camp. Afterward the children were asked questions about it, including questions about events that did not happen in the film. One such question was, "What did the boy say Sullivan had stolen?" when in fact no thefts had taken place in the film. The children were pressed to give some sort of answer, and the experimenters often suggested an answer. When the children were interviewed a week later, the false events as well as the suggested answers had been incorporated into their recollections of the movie. These false memories are discussed further in chapter 3.

Subjects of hypnosis may confabulate when they are asked to recall information associated with crimes (Dywan 1995, 1998), causing researchers to warn criminologists about the dangers of obtaining information from hypnotized subjects. There are also anecdotal reports of hypnotized subjects confabulating when asked why they did something in accord with their hypnotic suggestion. For instance, a stage hypnotist gives his subject the suggestion that he will wave his hands whenever he hears the word "money." When asked later why is he is waving his hands, the subject replies "Oh, I just felt like stretching."

In addition to the experiment described earlier, in which shoppers were asked their preferences about nylon stockings, Nisbett and colleagues conducted several other experiments that also seem to show confabulation by normal people. In one study, subjects with insomnia were given a placebo pill and were told it would produce rapid heart rate, breathing irregularities, bodily warmth, and alertness–all normal symptoms of insomnia (Storms and Nisbett 1970). The experimenters' idea was that knowledge of having taken the pill would cause the subjects to fall asleep earlier because they would be able to attribute their symptoms to the pill, rather than to whatever emotional turmoil in their lives was actually producing the insomnia. This is exactly what happened; the subjects reported going to sleep more quickly. However, when asked why they fell asleep earlier, they seemed to confabulate: "Arousal subjects typically replied that they usually found it easier to get to sleep later in the week, or that they had taken an exam that had worried them but had done well on it and could now relax, or that problems with a roommate or girlfriend seemed on their way to resolution" (Nisbett and Wilson 1977, 238).

Philosophers have found the concept of confabulation useful in describing peoples' reports of what went on in their minds during certain tasks such as solving puzzles (see also Nisbett and Wilson 1977). Philosophers of a behaviorist bent find the idea of confabulation amenable to their contention that we do not have reliable access to what goes on in our minds; that is, that introspection is not to be understood on the model of seeing, and that reports of introspections are not similar to reports of seen events. Dennett says, "there are circumstances in which people are just wrong about what they are doing and how they are doing it. It is not that they lie in the experimental situation but that they confabulate; they fill in gaps, guess, speculate, mistake theorizing for observing.... They don't have any way of 'seeing' (with an inner eye, presumably) the processes that govern their assertions, but that doesn't stop them from having heartfelt opinions to express" (1991, 94).

These normal people seem similar to the split-brain patients described earlier, who confabulated about why their left hands performed certain actions. In chapter 7 we examine the idea that certain reports of mental states are confabulations.

1.3   Features of Confabulation

Two very different sorts of activities might equally well be described as defining confabulation. The first activity has as its goal defining the word based on how people use it; the second focuses on determining what confabulation itself is. To put it another way, the first activity describes what the word *means* to people who use it, and the second focuses on what the word *refers* to.

The first activity involves discovering what criteria or conditions people normally apply when they use a word. The characteristics that we apply for "gold," for instance, include a yellowish color and a somewhat soft and heavy consistency compared with other metals. What actually makes a metal gold, however, is that it is made up of a certain type of atom; it has an atomic weight of 79. Some substances may look and feel like gold, such as iron pyrite (fool's gold), but are not gold because they lack the atomic and molecular structure of true gold (Putnam 1971; Kripke 1977).

Because of the distinction between the meaning and referent of a word, people are sometimes wrong about what they are referring to. People thought that "jade," for instance, referred to a single type of green stone, which makes elegant jewelry. It turns out that they were wrong. There are two chemically different types of stone to which this term is applied, one of which is much more attractive (jadeite) and valuable than the other (nephrite). Closer to our topic, "schizophrenia," for example, may turn out to refer to more than one type of brain disorder (Jeste et al. 1996).

The second activity called defining confabulation involves delineating the actual phenomena that people refer to when they use the word, even if they are wrong in their beliefs about them. Each of the seven criteria discussed in the following pages can initially be seen either as part of the meaning of the term "confabulation," or as designating its actual features. As with jade and gold, these two aspects can be teased apart, which we will have to do on certain occasions. We might agree that a particular criterion is currently part of the meaning of "confabulation", but find on investigation that it is not actually a feature of the phenomenon itself once we understand what confabulation actually is. These criteria and their proposed features are discussed next.

Criteria for Confabulation

1. Does the Patient Intend to Deceive? The orthodox position is that the patient has no intent to deceive. Ramachandran's result, in which patients who denied their paralysis attempted two-handed tasks, is one piece of evidence in support of this claim. Confabulation is not lying, which involves clear intent to deceive.

I lie to you when (and only when)

1.   I claim p to you.
2. p is false.
3.   I believe that p is false.
4.   I intend to cause you to believe p is true by claiming that p is true.

Confabulators do not satisfy the third condition since they seem to steadfastly believe what they say. The following dialogue, from DeLuca (2001, 121), shows a man with an anterior communicating artery aneurysm who is sincere about what he claims:

Doctor: You indicated last night you were working on a number of projects at home.... What would you say if I told you you were actually here in the hospital last night?
Patient: I'd be surprised, because my experience, what I learn from my eyes and ears tells me differently.... I'd want some evidence. I'd want some indication that you knew about my private world before I gave any cognizance.
Doctor: Would you believe me?
Patient: Not out of the blue, especially since we haven't even met (an illustration of the patient's amnesia).
Doctor: What if your wife was here and she agreed with me, what would you think at that point?
Patient: I'd continue to resist, but it would become more difficult.

2.   Does the Patient Have Some Motive Behind His or Her Response?
This criterion is of course related to the first one about intent to deceive; lying typically functions in the service of some motive. The most obvious motive in the case of confabulation would be a desire to cover up one's deficit. Some older textbooks combine the questions of motive and deception in a way that looks as if the authors are claiming that deception is intentional. Freedman et al. said that the patient recovering from amnesia after head injury "usually has a tendency to confabulate in order to cover his recent memory defect" (1975, 1428). Whitlock (1981) took these authors to task for their apparent attribution of intention to deceive, but the authors may not actually have been guilty of this. They were stumbling over a problem philosophers have encountered in their investigation of the problem of self-deception—how to describe behavior that seems to fall between the intentional-unintentional distinction. We will pursue this question in detail in chapters 8 and 9.

Another motive sometimes cited is the desire to avoid something known as the "catastrophic reaction," in which the patient comes to the horrible realization that he or she is paralyzed, has lost her memory, etc. and becomes depressed (Gainotti 1975; Zangwill 1953). A third motive that may also be at work in normal cases of confabulation is simply the desire to avoid saying, "I don't know," especially when the provoking question touches on something people are normally expected to know.

3.   Must a Defective Memory Be Involved?   Because it was first encountered in memory disorders, confabulation is traditionally defined in such a way that it must be accompanied by a memory deficit. Berlyne's classic definition of confabulation is that it is "a falsification of memory occurring in clear consciousness in association with an organically derived amnesia" (1972, 38). Similarly, Mercer et al. stated that "a necessary (though not sufficient) prerequisite for confabulation is impaired memory function" (1977, 433). Patients with Capgras' syndrome, however, do not have an obvious memory deficit. Nathaniel-James and Frith (1996) also argued that their schizophrenic patients exhibit confabulation in the absence of a memory deficit (see also Weinstein et al. 1956; Weinstein 1996). Similarly, confabulation in split-brain patients and in anosognosics is not accompanied by any obvious memory deficit. Even in the case of the memory confabulation syndromes, such as Korsakoff's, we will see that there are several indications that amnesia and confabulation are caused by damage to different brain areas.

4.   Must the Confabulation Be in Response to a Question or Request?
The traditional approach is of course covered by the distinction between *spontaneous* and *provoked* confabulation. This difference is important, since the questioning of the examiner sets up a special context in which an authority figure is soliciting information, which is not present in the case of spontaneous confabulation. Several outcomes seem available here. It might turn out that spontaneous confabulation is not actually confabulation, but simply the expression of a delusion. Alternatively, it might turn out that there are two types of confabulation, and the spontaneous-provoked distinction does not draw the correct boundary between the two types.

5.   Does the Confabulation Fill a Gap?   According to this criterion, confabulations fill in gaps at a certain level in the cognitive system. Perhaps this is because confabulation is another example of a tendency exhibited at many levels of brain function, to produce complete, coherent representations of the world. There are, however, several problems with Korsakoff's idea that confabulation fills a gap in the patient's memory. For one thing, a patient with Korsakoff's syndrome does not merely have a gap in episodic memory about recent activities; there is nothing there at all. The "gap" is not properly filled either, since it is filled with a (probably) false claim, with mud rather than mortar. It might seem *to the patient*, however, that he has merely a gap in his memory, and that confabulation fills that gap; but again, this may be making it all too intentional.

6.   Are Confabulations Necessarily in Linguistic Form?   Several researchers have categorized nonlinguistic responses as confabulatory. Lu et al. (1997) had patients point to fabric samples with one hand to indicate which texture of fabric they had been stimulated with on the other hand. The patients also had the option of pointing to a question mark in trials in which they had not been stimulated, a nonlinguistic version of answering "I don't know." The authors operationally defined confabulation as failure to point to a question mark. Bender et al. (1916) applied the term "confabulation" to the behavior of patients when they produced meaningless drawings as if they were familiar designs. Similarly, Joslyn et al. (1978) had patients reproduce from memory certain drawings they had seen, and described cases in which the patients added extra features that were not actually present as confabulations (see also the work of Kern et al. [1992] on Alzheimer's disease, and Chatterjee [1995] on neglect).

These uses of confabulation seem to conflict with the idea that confabulators tell stories, which are usually false. Pointing at a piece of cloth, or answering "yes" rather than "I don't know" to a question do not seem to be confabulations according to this view. Those researchers seemed to be conceiving of confabulation as a broader epistemic phenomenon, rather than as a narrower, purely linguistic one. In the epistemic view, a confabulation is a poorly grounded claim, and a confabulatory person tends to make epistemically ill-grounded claims. Similarly, "to confabulate," means something like "to confidently claim something one has a poor epistemic warrant for."

For confabulations that are linguistic, a further question is whether they must be internally consistent. Weinstein (1996) argued that confabulations should be coherent, as did Talland (1961). Contrary to this, Moscovitch (1995), stated that they "need not be coherent and internally consistent"; there are several examples of confabulation, even of the provoked variety, in the literature that contain contradictions. Another semantic question about confabulations that are responses to questions is whether they all must be false. A confabulatory patient may on occasion say something true, perhaps by accident.

7.   Are Confabulations the Result of Delusions?   That is, what is the relation between confabulation and delusion? It would be wrong to classify even spontaneous confabulation as a type of delusion, since delusions are, minimally, false or ill-grounded beliefs (but see Stephens and Graham 2004), whereas confabulations are false (or ill-grounded) claims. (Similar questions arise about the wisdom of making falsity part of the definition of delusion, as arise in the case of confabulation; see Fulford 1989.) A delusion might give rise to a confabulation, however, which seems to be what happens in the case of Capgras' syndrome. One way in which confabulation was differentiated from the expression of a delusion in the past was by the claim that confabulation necessarily involves a memory deficit (Berrios 1998). However, with the addition of several syndromes that do not involve memory deficits to the list of confabulation syndromes, this criterion may have lost its usefulness. Delusions also tend to be long lasting; they are described by the American Psychiatric Association (1994, 765) as "firmly sustained in spite of what almost everyone else believes and in spite of what constitutes incontrovertible and obvious proof or evidence to the contrary." Many confabulations, on the other hand, are quickly forgotten after they are communicated. Confabulators do tend to resist changing their claims in the face of contrary evidence, but not with the tenacity of the deluded.

In chapter 8, after we have surveyed the data on confabulation, we will construct a definition of confabulation. There we will examine several of these possible criteria for inclusion in the definition.

1.4   Three Concepts of Confabulation

Once the different possible features of confabulation are described, it becomes clear that there are competing concepts of confabulation, each of which makes different sets of the features described here essential to the definition. We might think of a concept as a bundle of such features, with the more essential features at the center and the less essential ones at the periphery. I intend *concept* to have a certain neutrality or hypothetical nature. A concept can crystallize into a definition, which has happened in the case of the mnemonic concept. Two other concepts seem to be at work in the minds of people using "confabulation"—the linguistic and the epistemic—but they have not yet been formally defined.

Mnemonic Concept

According to the classic concept, confabulations are stories produced to cover gaps in memory. Memory and gap-filling features are essential in this definition. Since its introduction at the beginning of the twentieth century, however, the concept has been increasingly threatened by applications of confabulation to patients who have no obvious memory problem. Because of its reliance on the notion of gap filling, the mnemonic concept also must address the question of whether the confabulator does this knowingly, with intent to deceive. Most people employing the classic concept agree that confabulators do not deliberately attempt to deceive; thus Moscovitch's (1989) description of confabulation as "honest lying."

Linguistic Concept

Here, confabulations are false stories or sentences. In this concept, confabulation is a symptom shared by a set of syndromes with diverse physical bases, in the way that high blood pressure is the name of a symptom with many different causes. The noun *confabulation* is defined first as a false narrative, and the verb *to confabulate* is the act of producing a confabulation, in the noun sense. This concept is truer to the linguistic meaning in the etymology of confabulation as an act of conversing, or talking with someone. Proponents of the linguistic concept might trace the roots of confabulation to the penchant we humans have for telling and listening to stories. Since this concept is neutral about the question of whether the person intentionally produced a false narrative, it can avoid the difficult question of intention that other concepts are saddled with.

One objection to the linguistic concept is that it causes us to mistakenly lump lying and confabulation together because both involve the production of false narratives. Its emphasis on the language itself also poses a special problem for this concept concerning the truth of confabulations. Are confabulations false by definition? Some of the memory claims of patients with Korsakoff's syndrome are true but misplaced in time. Similarly, even a highly confabulatory patient will occasionally make true claims. The obvious way around this, to note that a patient might happen to say something true but that it would be a matter of luck, pushes the linguistic concept closer to the epistemic concept. On the other hand, if different syndromes that result in confabulation are not found to share any significant physical basis, this will lend support to the linguistic concept; all they have in common is a symptom: the production of false narratives.

Epistemic Concept

In this concept, a confabulation is a certain type of epistemically ill-grounded claim that the confabulator does not know is ill-grounded. The claims need not be made in full natural language sentences; they may consist of drawings, pointing to a picture, or simple "yes," and "no" answers to questions. Dennett's use of confabulation, quoted earlier, "It is not that they lie in the experimental situation but that they confabulate; they fill in gaps, guess, speculate, mistake theorizing for observing," (1991, 94) seems to have the epistemic concept behind it. Studies cited earlier that counted pointing to a cloth sample or producing certain drawings as confabulation also may be employing an epistemic concept. Both activities involve making a claim, the claim that I was stimulated with *this* type of texture, or *this* is the drawing I saw.

One way to delineate the epistemic concept further would be in terms of certain malfunctions of normal belief forming and expressing processes (Goldman 1986). The malfunctioning process in the case of confabulation may be one that allows people to attach doubt to ideas; con-fabulatory people do not experience doubt about their claims and ideas, whereas a normal person would. The claims are epistemically ill-grounded because they have not passed a review process that can result in doubt about them. That process does not function correctly because of brain damage. This is one way to explain the connection between confabulation and the disinhibition that frequently accompanies it; the doubt normal people feel when they consider certain ideas is caused by the same process that normally inhibits inappropriate responses.

The groundedness of our claims seems to come in continuously vari-able degrees, over which we normally have some voluntary control. We can loosen or tighten our epistemic criteria, depending on the situation. When people give legal testimony, for instance, they are able to apply abnormally high standards to what they say, much higher than when they are speaking casually with friends. We also take this sort of care in some everyday situa-tions: a man warns his friend before they meet with someone, "Be careful what you say to him." Have confabulating patients perhaps lost this ability to regulate the level of certainty each context requires?

By understanding confabulation as arising from malfunctioning epistemic processes, this concept is committed to going beyond the ex-pression of confabulation to the processes behind it. One objection to this concept is that it applies confabulation too broadly and hence reduces the likelihood of our finding a single neural basis for it. The way to respond is presumably to argue in favor of a malfunction of a general, high-level brain process, one that is involved in all the syndromes of confabulation.

1.5   Mirror-Image Syndromes

Sometimes one can understand something better by asking what the op-posite of that thing is. Constructing a system of opposites also helps to provide a large conceptual space in which to place confabulation. The existence of such opposite or mirror-image syndromes also helps to dispel any notion that confabulation is somehow a universal feature of any sort of serious brain damage. Of course, there is no such thing as *the* opposite of something; rather, something is the opposite of another thing in a certain respect. One feature of confabulation is the confabulator's claim that he has certain abilities that he in fact lacks, such as the ability to remember, to move his arm, or, in the case of split-brain patients, to explain why his left arm did something. In contrast with this, there is a class of neuro-logical syndromes in which patients *do* possess certain abilities, but claim that they do not. *Blindsight* patients, for example, have a large blind spot in their visual fields that is due to an injury, and will claim that they cannot see anything at all there (Weiskrantz 1986). If they are asked which way a beam of light moved around in the blind area, they are able to answer correctly because other, nonconscious streams of visual processing are intact. In a case even more closely related to confabulations about vision, Hartmann et al. (1991) described a patient with cortical blindness who suffered from what they called "inverse Anton's syndrome." The man claimed to have no ability to see at all, yet testing revealed that he had a small area of the visual field in which his visual abilities were well preserved, including the ability to name objects, colors, and famous faces, as well as to read words.

Similarly, some patients with *prosopagnosia* (*face* agnosia), who claim to be unable to visually recognize familiar people, show unconscious autonomic reactions (as measured by activity in the sweat glands of their hands) to photographs of people they know (Bauer 1984; Tranel and Damasio 1985), just as normal people do. Bauer (1984, 1986) and Ellis and Young (1990) hypothesized that prosopagnosics suffer from a syndrome that is the mirror image of Capgras' syndrome, suggesting that Capgras' patients recognize relatives (albeit as impostors) but fail to register the normal skin-conductance response to the sight of them, a prediction that has been confirmed experimentally (Ellis et al. 1997; Hirstein and Ramachandran 1997). We examine the distinction between prosopagnosia and Capgras' syndrome in greater detail in chapter 5.

I noted that another feature of confabulation seems to be an inability to doubt what one says; it involves a sort of pathological certainty. The opposite of this would be pathological doubt. Two sorts of conditions are possible opposites in this regard: paranoia and obsessive-compulsive disorder (OCD). Paranoia may not be a good candidate, since it has been observed to coexist with confabulation (Berlyne 1972; Weinstein 1996), and one might contend that it involves more suspicion than doubt. Obsessive-compulsive disorder, however, may be a good candidate. It can be interpreted as pathological doubt—doubt that one's hands are clean, or that all the doors are locked, or that the stove has been turned off, for example (Schwartz 1998; Saxena et al. 1998). A possible explanation of the relation between confabulation and OCD is that the process that produces a feeling of doubt and is hyperfunctioning in OCD is completely broken in confabulatory patients (see chapter 4).

1.6   Conclusion: Setting the Problem of Confabulation

The problem of confabulation is one of giving a satisfactory explanation for what exactly the phenomenon is, and what it tells us about who we are, how we think, and how the brain works. An account of confabulation should be able to answer the following questions:

1.   Do all behaviors referred to as confabulations involve the same brain malfunction?

a.   What are the important features of the phenomenon (or phenomena if it is not unitary)?

b.   What causes confabulation?

c.   Why does confabulation appear in so many disorders?

d.   Are there different subtypes of confabulation? What are they?

e.   What is the connection between confabulation and denial of illness?

f.   Why are some disabilities admitted and others denied?

g.   How can confabulation be prevented or treated?

2.   What does the existence of confabulation tell us about human nature?

a.   Is confabulation strictly a pathological phenomenon, or is there a continuum, shading from pathological cases into normal behavior?

b.   What is the connection between confabulation and the self?

c.   What is the connection between confabulation and self-deception? Are statements made by self-deceived people confabulations?

d.   Does confabulation have positive functions?

Obviously, those adhering to different concepts—mnemonic, linguistic, and epistemic—have different answers to many of these questions. A comparison of how the concepts handle the phenomena of confabulation will prove informative in what follows. But more important, the interplay between the concepts and the empirical evidence will, I hope, lead to our understanding of what this fascinating phenomenon is. In the ensuing chapters, I propose that the epistemic concept is preferable, because it nicely captures a natural set of the phenomena of confabulation, but also because it provides a productive and substantially correct guide to investigating and understanding the brain malfunctions behind confabulation.

2 Philosophy and Neuroscience

If ... it is not in my power to arrive at the knowledge of the truth, I may at least do what is in my power, and with firm purpose avoid giving credence to any false thing.
—René Descartes (1641)

2.1 The Growth of Neuroscience

The roots of the problem of confabulation are spread broadly across several different disciplines, including biology, psychology, and philosophy. The roots are also deep historically, in that reports of confabulation go back thousands of years. In the early 1900s neurologists formed the first concept of confabulation, and neurology has been the primary field investigating the phenomenon up to the present. This fact will give chapters 3 through 7 a distinctly neurological character, as we examine each syndrome. As soon as neurologists postulated the existence of confabulation, they began attributing it to localized brain lesions. In the early days of neurology—the middle to late 1800s—the science relied heavily on the raw genius of people such as John Hughlings Jackson, and Jean-Martin Charcot, a tradition that continued into the 1950s, 1960s, and 1970s with neuroscientist-writers such as Paul MacLean, Alexander Luria, and Norman Geschwind. These individuals had an intuitive ability to add up all the information about a condition or syndrome and point to a spot in the brain that might be the site of damage. Now, however, research into how the brain works at all levels has caught up with these intrepid explorers. Instead of shots in the dark, comprehensive theories are under development, supported by converging evidence from each of the neuroscientific subdisciplines; there is increasingly less darkness in which to shoot. Today the work of the early neurologists has been taken up by new fields they helped to develop, such as cognitive neuroscience, and especially cognitive neuropsychology.

Neurologists have always used the brain itself as the common grid on which different theories of confabulation can be overlaid, compared, and tested. Their ultimate goal was to be able to point to a specific problem area in the brain for any given neurological syndrome. They learned quite early on that it was not going to be that simple, but the basic drive for localization lives on in various modified forms. Knowledge about lesion areas known to cause confabulation has had to wait until the structure and function of these areas was understood before scientists could begin posing their theories of confabulation in detailed and testable forms, something that has happened only within the past thirty or so years.

One of our overall goals is to merge this neurological knowledge with philosophical analyses to obtain a satisfying explanation of the basic phenomenon of confabulation. A complete understanding of the implications of confabulation would involve forays into perhaps all of the humanities; into how it figures in writing, both fiction and nonfiction; but also political science, through examination of the confabulatory behavior of political leaders; in economics and decision theory; and so on. In terms of a basic description of the phenomenon and how it figures in our mental lives, however, neuroscience possesses the best-developed theories and concepts. We can learn facts about confabulation from neurobiology that would have been difficult if not impossible to learn by alternative techniques, such as an analysis of confabulatory speech behavior, no matter how intensive. Psychology provides a valuable paradigm here also; the sorts of behaviors that constitute confabulatory acts are traditionally within its subject matter. Finally, a full understanding of confabulation requires techniques and information from philosophy, including information about the nature of knowledge and techniques used for constructing definitions. These techniques will prove useful in chapters 8 and 9, where I attempt a definition of confabulation and compare it with definitions of "deceive," "lie," and "self-deception." This loan from philosophy is quickly repaid. Confabulation can shed light on several classic philosophical problems, including questions about human knowledge seeking and self-deception.

Or so I maintain. Since my hope is to speak to readers from at least those disciplines, I will try to describe things in a way that is accessible to all. However, differences in the methods of analysis, style of argument, and standards of proof between neuroscience and philosophy may lead to misunderstandings. One must take care to avoid mistaking violations of disciplinary conventions for genuine mistakes about the subject matter. This was not always a problem; these sorts of disciplinary boundaries are recent and largely arbitrary. A hundred fifty years ago it was *all* philosophy; what we call science today was then called natural philosophy. Psychology became a discipline separate from philosophy only around 1880. Whereas these boundaries are arbitrary, they have unfortunately caused very real gaps in understanding. There is cause for optimism, though. The many successful multidisciplinary endeavors that exist today militates strongly against making boundaries too rigid. After all, anyone genuinely interested in solving a certain problem has to follow the clues wherever they lead, without regard to which discipline's turf they may lead. One is at risk of looking like a dilettante, but it is a risk worth taking if we really want to learn something.

Both neuroscience and philosophy are cognitive sciences; the others include computer science (particularly artificial intelligence), linguistics, psychology, and anthropology. A healthy competition for the heart of cognitive science exists between the traditional cognitivist approach centered in cognitive psychology, and a growing approach based on neuroscience. Neuroscience has always been considered part of cognitive science, but the base traditionally resided in cognitive psychology and in the traditional computer modeling approaches of the artificial intelligence theorists. Recently though, the neuroscience-centered approach has attracted followers with its impressive brain-imaging techniques and fascinating clinical case histories, which can now include much more accurate hypotheses about which brain areas and functions are damaged.

Cognitive neuroscience arose as a subfield in the late 1970s, when language researchers such as neurologists examining aphasias, linguists constructing grammars, and computer scientists modeling human language use realized they were climbing different sides of the same mountain. Along with the more traditional and more popular case histories of neurologist Oliver Sacks (1998), the neuroscience-based approach has attracted attention with several accessible and revolutionary books emerging primarily from cognitive neuroscience, such as Antonio Damasio's *Descartes' Error* (1994) and Joseph LeDoux's *Synaptic Self* (2002).

The confluence of findings in neuroscience about the anatomical, physiological, and chemical properties of the brain, allows checking, comparison, and testing along hundreds of different connecting threads. Both neuroscientists and philosophers should continue to explore and try out these theoretical frameworks in such problems as understanding how behavior is produced, or how our mental lives are realized. At the highest level, this is a process of arranging groups of consistent theories from several disciplines, as I will sometimes do here. This metatheoretical process often employs theories and hypotheses as its data, and involves a search for trends in what is hypothesized as well as what is well confirmed. Consistencies and inconsistencies are noted, for instance between what neurologists say a part of the brain does and what neuroanatomists say. More specifically, I will look at the possibilities and problems involved in synthesizing some of the hypotheses and theories, which are for the most part designed to explain only one or two confabulation syndromes, into a larger theory of confabulation in general. The objective is to find a consistent and explanatorily powerful mix of theories.

Our focus within neuroscience will be at the level closest to our phenomena as they are described, and the one in which the confabulation literature has been posed: gross anatomy and the upper or more general levels of neuroanatomy and neurophysiology. This branch of neuroscience looks at large systems, involving multiple cortical areas and subcortical nuclei. It is the most natural and intuitive way to approach the brain, as the first anatomists did when they examined the strangely shaped brain components and nuclei, and wondered, "What does *this* do?" Although a great deal remains to be understood, we are beginning to see much more clearly how brain anatomy connects to psychology. Contemporary neuroscience has reached a threshold moment when progress increases exponentially, for a number of reasons.

Advances in Neuroscience

Brain Imaging   Several kinds of brain-imaging techniques now exist, such as classic radiographs and computed tomography scans, in addition to the more recent positron emission tomography, magnetic resonance imaging (MRI), and single photon emission tomography; many other techniques are under development. There are also the classic techniques of brain stimulation, from simply applying electricity to exposed brains as Penfield did in the 1950s, to more recent, less invasive techniques such as transcranial magnetic stimulation. Each imaging technique has a set of brain properties to which it is sensitive and a much larger set to which it is blind. By combining techniques, for instance by using MRI and functional MRI (fMRI), differences among individual subjects can be greatly reduced, allowing more precise localization of functions.

One must always keep in mind the limitations of each technique, however. Most imaging methods measure the amount of energy the brain is using, but this leaves them vulnerable to missing events that are important but do not involve a great expenditure of energy. Rolls, for instance, issued a warning about imaging results directed at a brain area that will prove of interest to us, the orbitofrontal cortex: "An imaging study might not reveal really what is happening in a brain structure such as the orbitofrontal cortex where quite small proportions of neurons respond to any particular condition; and, especially one would need to be very careful not to place much weight on a failure to find activation in a particular task, as the proportion of neurons responding may be small" (Rolls 1999, 121).

Anatomical and Physiological Investigation Techniques   Neurobiologists have found it useful to speak of brain *circuits*. These typically consist of a set of interconnected areas that modify each other's activity in certain ways. In the past twenty-five years we have learned a great deal about where these circuits are and how they are connected. One of the most useful techniques for obtaining this sort of knowledge is a cell-staining study in which a colored chemical dye that naturally follows neuronal connections is used to determine which brain areas the dyed area causally interacts with.

Cell-recording techniques involve inserting sensitive electrodes that can detect the firing of individual neurons. Once the electrode is in place, researchers expose the subject to different stimuli or induce different behaviors, as they measure the activity of the neurons. In humans, this technique is used with epileptic patients to determine what functions might be compromised by surgery to remove a brain area from which seizures originate. These methods are augmented by a wide variety of behavioral tests in attempts to discern the particular functions that each circuit performs. The cognitive neuropsychologist relates what is known about brain anatomy to the behavioral evidence. For example, what abilities does damage to this part of the brain compromise, and what does this tell us about how the brain works? It often requires great creativity to find the exact ability that a certain type of brain damage affects.

Neurocomputation Computational neuroscience is helping us understand how neurons perform different sorts of computations. Characterizing the electrochemical interaction taking place between neurons with a computational theory has proved difficult. Bias based on overapplication of the computer metaphor to a very uncomputerlike biological system produced an initial tendency to think that a neuron's firing or not firing corresponded to a computer's 1s and 0s. Neurons have proved to be far more complicated than this. Neurocomputationists are fully past this problem, however, and now focus on producing realistic computer simulations of brain processes based on a close reading of what the neuroanatomists are saying (see Churchland and Sejnowski 1992).

Relating Lesions to Functions

It is also important to find out which brain circuits embody or support conscious states, and to discover the different functions these circuits have in our conscious mental lives. As the functions of the higher brain organs, networks, and circuits are understood and described, the hope is that we can begin to see the underpinnings of our everyday mental states. However, the crucial information needed to do this is often missing, in that much less is known about which brain circuits match up with the sorts of mental events we are aware of every day. For instance, what happens in the brain when we feel emotions? Which parts of the brain do we think with? Where are mental images located? Where is consciousness itself located? Here the cognitive neuropsychologist struggles to keep up with the anatomists and physiologists. Damage to the brain wipes out entire circuits while leaving others intact. More typically, some part of the brain is destroyed by internal forces (stroke, tumor, alcohol, virus) or external ones (a gunshot, the surgeon's scalpel).

Cognitive neuropsychologists attempt to understand the relation between brain damage and behavior. For their purposes, the more focal the patient's brain damage the better. Patients with diffuse brain damage are not good subjects because it can be difficult to determine which damaged area is causing a given behavior (or lack of one). Even if the damage is restricted to a single, spatially defined area, most areas of the brain participate in more than one circuit, and hence it is rare that damage disrupts only a single circuit. In addition, other circuits can take up the tasks formerly accomplished by the damaged one. This is good for the patient of course, but it makes the theorist's task much more difficult.

There is a long history of battle between localizationists, who believe that brain functions can be assigned to discrete brain circuits or areas, and holists, who believe that this is not realistic, given the interconnected nature of the brain. Of course, not even the most extreme localizationists believe the following maxim: If damage to area x causes an inability to do y, this implies that the function of area x is to give the organism the ability to do y. Stated in this general way, the maxim is false; nevertheless, sometimes both the following propositions are true of a certain area x and function y: (1) Damage to area x causes an inability to do y and (2) the function of area x is to give the organism the ability to do y. For instance, severe enough damage to both eyes causes inability to see, and it is true that the function of the eyes is to give an organism the ability to see. As usual, the truth lies somewhere between extreme localizationism and extreme holism, as the doctrine of brain circuits suggests. In this approach, brain functions are performed by interconnected systems of discrete brain areas.

Understanding Connections between Areas

It is a basic assumption of neurologists that the function of their discipline's intrinsic link to biology is to make their theories as explicit and detailed as possible. However, only recently has biology developed the information necessary to make this a real option. Neurological programs of inquiry are sometimes criticized for claiming merely that brain areas x and y are connected. But as a skeptical friend once said to me, if the question is whether any given brain areas x and y are connected, the answer is "yes." Merely saying x and y are connected is not interesting given the brain's vast interconnections. No one is claiming that any two cortical areas are directly (or "monosynaptically") connected, but among many issues this skepticism raises is that even the notion of being directly connected requires a great deal of clarification. Ultimately the following questions and more about the nature of the connection have to be answered:

1.   Is the connection one way, or is it reciprocal?
2.   What sorts of cells are the connecting fibers themselves?
3.   What cortical layers do the connecting cells originate from and terminate in?
4.   How dense is the connection?
5.   What neurotransmitters mediate activity across the connection?

The existence of connections between two brain areas or organs is important because it indicates the presence of processes carrying causal influences from one area to another (and sometimes back). The brain follows evolution's general use-it-or-lose-it strategy; existing connections are there for a reason.

2.2   Principles of Brain Structure and Function

Although it is important to relate findings in neuroscience to our everyday ways of thinking about the brain, we must be open to the possibility that the brain divides the set of mental functions in quite different ways from our everyday categories. Anatomists are much more comfortable reporting on brain structure than speculating about function, and they seem to do the latter only grudgingly. Nevertheless, these speculations are of great interest for a couple of reasons. First, they represent a special perspective on the phenomena of interest. The anatomist studies the processes all cognitive scientists are interested in—thought, memory, and so on—but from a completely different angle. Second, these speculations are based on findings that are more concrete than the data of the other cognitive sciences.

Often anatomists' discoveries about structure and function were obtained through the study of animals other than humans, usually rhesus monkeys or white rats, but sometimes dogs, cats, orangutans, or owls. The implications for human behavior can be difficult to discern; for instance, what do findings about the feeding behavior of rats have to do with humans? However, all of these data are of use. It is a truism of neurobiology that throughout our evolution the brain grew outward from the brainstem by adding and enlarging structures to what already existed, rather than replacing or eliminating existing structures. The cortex, the wrinkled, gray outer covering of the brain, is its most recent evolutionary addition and the place where representation and higher-level cognition—perception, thought, intention, and so on—are believed to take place. The brain's evolutionary history gives it a multilayered structure, and because we share much of this structure with other species, the human nervous system must be understood in the context of the nervous systems of related animal species, especially higher primates.

Those interested in certain parts of the brain, we for instance, with our interest in areas destroyed in confabulation syndromes, can examine and weigh different claims about the functions of these areas. But of course we do not understand the functions of areas, circuits, or organs in isolation; we understand them as part of a larger functional system. One knows precious little about the fuel pump in a car if all one knows is that it pumps fuel. Where does the fuel come from? Where does it go? How does the pump contribute to the larger goal of moving the car down the road?

When one is unsure about the function of a given area or circuit, however, theories about the functions of areas connected to it can also provide valuable information for testing and improving your hypotheses. What one needs ideally to understand a brain area is a *functional map*, knowledge of both that area's functions and the functions of its connected areas.

One fundamental functional unit of the brain is the *perception-action cycle*, a continuous chain of processes linking perception, cognition, and intention, and ending in action. "Sensory information is neurally processed, and the result of this processing is converted into movement, which induces changes in the environment (both internal and external). These changes, in turn, generate new sensory input, leading to new movement and that to more changes and so on" (Fuster 1995, 274). These outward-reaching circuits are accomplished by dozens of inner circuits running between different brain organs and areas. Having multiple layers of perception-action cycles is good engineering; it creates a nervous system that is capable of keeping the organism alive despite all types of damage, and puts emphasis on the minimal unit required to produce actions informed by perception.

John Hughlings Jackson (1881, 1884) saw early on that the nervous system had this sort of structure: many levels, with perception and action at every level. He also realized another effect of damage to such a design: interruption of higher levels might have a releasing effect on the action component of lower levels. This phenomenon has come to be called *disinhibition*. Since the frontal lobes contain parts of many higher-level circuits, this idea suggests that frontal damage will have such a releasing effect. During the twentieth century, experiments with animals and unintentional experiments on people, such as frontal lobotomies, bore this out. "There is considerable agreement that, regardless of the species (cat, rat, monkey), frontal ablation does not drastically change the premorbid behavior; instead the prior behavior becomes accentuated" (Stuss and Benson 1986, 126). The notion of releasing implies that one brain state holds another in check, or inhibits it.

This seems to be another principle of brain function: The brain likes to balance systems against one another; for instance, the way that it balances the sympathetic and parasympathetic branches of the autonomic nervous system against each other. One clear engineering benefit of such an arrangement is that it allows precise adjustments in either direction. To give a concrete example, cars have one system that makes them go faster, the engine and drive train, and another, opposed system that makes them go slower, the braking system. One could try to combine both of these in a single process, but our current divided setup has proved to be far more effective and simple. Systems of opposed processes have a fatal flaw, however. When one of the processes is damaged, the other one can run out of control, producing highly undesirable results.

Another truism connecting structure to function in the brain is that, especially within the cortex, perception tends to begin at the back of the brain. Then, as the perceptual products are refined, analyzed, and employed to plan and guide behavior, the processing moves toward the front of the brain. Thus the general direction of processing flow in the cortex is from back to front. This movement occurs along several parallel processing streams, each with its own special representational domain. Processing streams more dorsally located (figure 2.1) in the cortex tend to represent the body and its nearby space, in contrast to processing in the more ventral areas of the cortex, which tend to represent objects and people of interest. As we will see in chapter 5, some of these latter streams are specialized for perception of faces. Because all circuits involved in the

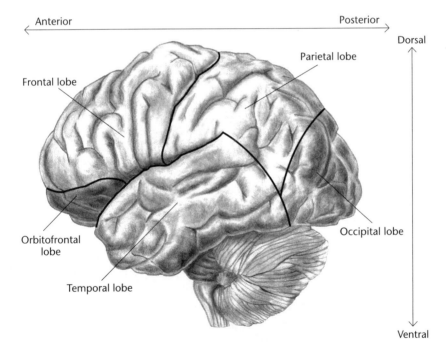

Figure 2.1
Lobes of the brain, lateral view. Visible are the frontal, orbitofrontal, parietal, temporal, and occipital lobes. The front of the brain (left in this diagram) is referred to as the anterior (or rostral) part, and the back is referred to as the posterior (or caudal) part. The top of the brain is referred to as the dorsal (or superior) part and the bottom as the ventral (or inferior) part.

brain's accomplishment of a perception-action cycle are often active simultaneously, it is impractical to look for the exact point where perception ends and action begins. We typically act and perceive at the same time; we perceive in order to act, but we also act in order to perceive. Thus, while it is true to say that there is a basic back-to-front movement trend in the cortex, the movement is not the steady bottom-up progression envisaged by cognitive psychologists of the 1960s and 1970s. There are many clear demonstrations of top-down effects in which higher levels of cognition influence lower levels of perception. Apparently the brain likes to employ both top-down and bottom-up processes simultaneously; expectation influences perception while perception modifies expectation.

Researchers have moved relentlessly inward from the brain's two areas of interface with the world, the sensory and motor areas, toward the unknown region in between, which lies largely in the frontal lobes (figure 2.2). The front of the brain is the last part to mature, in adolescence (Case 1992; Smith et al. 1992), and it is among the last parts of the brain for which we have been able to develop good theories. The first parts of the cortex we were able to understand are divided into different areas that

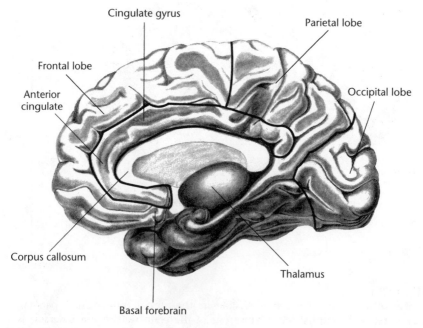

Figure 2.2
Lobes of the brain, medial view. Visible are the frontal, parietal, temporal, and occipital lobes. Also visible are the cingulate gyrus, corpus callosum, and thalamus.

function as *topographic* maps in one of the sense modalities. A topographic map mirrors some property of the mapped events. Standard road maps, for instance, mirror the spatial relations between the locations on the map: Baltimore is between Philadelphia and Washington on the map because that is the actual spatial relation of those cities. In the brain, vision is accomplished in part by a linked set of *retinotopic* maps, areas that preserve among themselves spatial relations between different rods and cones of the retina. Hearing is accomplished in part by a connected set of several *tonotopic* maps. Here the mirrored relations are the frequencies of incoming vibrations. Note D comes between notes C and E, and when a person hears those notes, D is (partly) represented by an area of the tonotopic map lying between the areas representing C and E. Incoming information is first processed in *unimodal* sensory areas. These areas tend to merge their inputs into *multimodal* areas that in turn project to higher-level areas in the prefrontal cortex.

Two Cortical Moieties: An Evolutionary-Functional Theory of the Cortex

The cortex can be divided based on the current configuration of its gyri and sulci, but these groupings are arbitrary and made strictly for spatial rather than functional or developmental reasons. Another approach is to use what is known about the evolutionary development of the cortex. In 1969, Friedrich Sanides said that if we look at the layers within the cortex, clear developmental trends can be seen. Anatomists speak of certain areas as "developing out of" other areas over the course of evolution. They determine this by studying the brains of increasingly complex species, and by using what is known about the brain's evolutionary history. This evidence points to the existence of two centers from which it is posited the entire cortex evolved. One center is in what is now the hippocampus, and is referred to as the archicortical, or dorsal cortical trend. It consists primarily of the upper and inner parts of the cortex. The other center is older in the evolutionary time scale. It originates at the posterior base of the orbitofrontal cortex (figure 2.1) and is called the paleocortical, or ventral trend. It consists mainly of the lower portions of the cortex. Processing in the two trends remains segregated even when they interconnect with the same subcortical organs. Pandya and Yeterian (2001) mention that the two trends are separate in the thalamic mediodorsal nucleus and the caudate nucleus (see chapter 3).

The ventral cortical trend originates at the posterior portion of the orbitofrontal cortex and gives rise to the ventrolateral prefrontal cortex (see chapter 4), the anterior insula, the inferior and lateral temporal lobes, including the multimodal areas in the superior temporal cortex (see chapters 4 and 5), and the inferior parietal lobe (see chapter 6). It also includes the somatosensory and motor areas that represent the head, face, and neck in ventral portions of the cortex near the central sulcus. The dorsal trend gives rise to the cingulate gyrus (see chapters 3 and 4), the parahippo-campal gyrus, the posterior insula, and the superior and medial parietal lobes. This trend contains somatotopic areas involved in representing the body trunk and limbs, as well as a posterior temporal region involved in auditory spatial localization, and "ventromedial temporal and dorsomedial occipital areas relating to the peripheral visual field" (Pandya and Yeterian 2001, 52). This trend also gives rise to the medial prefrontal regions and premotor regions in the dorsal cortex.

The ventral and dorsal trends each have their own visual system, but there are important differences between the two (Mishkin et al. 1983; Van Essen and Maunsell 1983). Each stream terminates in a visual "working memory area" in the lateral prefrontal cortex (Goldman-Rakic 1987; see chapter 3). The ventral visual stream specializes in recognition of objects, hence it has been called the *what* stream. Color vision is highly accu-rate in this stream, and damage here can remove color from the patient's vision (*achromatopsia*). Damage here can also destroy the patient's face-recognition abilities (see chapter 5). It is more correct to speak of ventral visual *streams*, plural. Two separate streams appear to be involved in per-ceiving other peoples' faces, one in determining their identities, and the other in identifying which emotions they are expressing.

The dorsal visual stream specializes in spatial vision, giving it its name, the *where* stream. Its primary use is in navigating around objects in an environment. Damage to the parietal areas that perform the computa-tions involved can produce disorders in reaching toward objects and in perception of motion, as well as disorientation in an environment.

2.3   The Limbic and Autonomic Systems

The Limbic System

As we will see, damage to systems known to be involved in producing cer-tain emotions has been implicated in causing confabulation. Tradition-ally, the role of providing the neural substrate for emotion was assigned to a set of interconnected brain organs, including the cingulate gyrus, amyg-dala, and hippocampus, together known as the *limbic system*. The idea that these organs truly constitute a unified and significant functional system in the brain has come under criticism, but the notion of the limbic system retains its currency in most circles. In 1937, Papez proposed that a set of brain structures that included the hippocampus, cingulate gyrus, parts of the thalamus, and other structures are responsible for emotions, including their conscious sensation (which he believed was due specifically to the cingulate gyrus). Papez attributed the visceral effects of emotions to a con-nected structure, the hypothalamus. MacLean (1954) expanded this system and for the first time referred to it as the limbic system. *Limbic* comes from the Latin word for border, and refers to the border formed around the brainstem by such parts of the cortex as the cingulate gyrus and hippocampus. Devinsky et al. (1995) distinguished the rostral and caudal limbic systems. The rostral system consists of the amygdala, septum, orbitofrontal cortex, anterior insula, anterior cingulate cortex, and ventral striatum (see chapters 3 and 4). The caudal limbic system consists of the hippocampus, posterior parietal lobe, posterior parahippocampal gyrus, posterior cingulate gyrus, and dorsal striatum.

The Autonomic System
Autonomic activity is crucial to emotion, and as we will see, reduced autonomic activity is found in several of the confabulation syndromes. The autonomic nervous system is made up of three subsystems: the sympathetic, parasympathetic, and enteric nervous systems. *Autonomic* means self-governing; the actions of this system are largely involuntary, although certain aspects of it can be brought under voluntary control through techniques such as biofeedback. The sympathetic and parasympathetic nervous systems have roughly opposing functions. The sympathetic system's function is to respond to emergencies by adjusting the person's internal environment, known as the fight or flight response. The parasympathetic nervous system maintains the internal environment in conditions of low stress, a function that is sometimes summarized by the phrase "rest and digest."

The connection between the autonomic nervous system and the limbic system is primarily through the hypothalamus, which is one of the main controllers of the autonomic system. Connections between the hypothalamus and the amygdala, and from there to the insular and cingulate cortices, are thought to play a role in higher-level activation and regulation of autonomic activity. The sympathetic nervous system is activated by the hypothalamus when a significant object or event is detected. Activation produces dilation of the pupils and the bronchi in the lungs, increased heart rate, constriction of most of the body's blood vessels, and secretion of sweat from the sweat glands. The parasympathetic system typically produces the opposite responses in the same organs: constriction of pupils, a reduced heart rate, dilation of blood vessels, and so on. We will discuss the autonomic system in more detail in chapter 4 when we examine one of its outputs, the skin-conductance response, used by polygraph operators in an attempt to detect lying.

2.4   Philosophy's Role

Some research in neuroscience has obvious application to philosophical issues. For instance, several attempts to locate the brain areas responsible for producing conscious states clearly are relevant to the classic mind-body problem. Our connections here are subtler, but my goal is to show that confabulation is relevant to philosophical questions about knowledge and questions about self-deception. Since my plan is to propose in chapter 8 that being confabulatory is to be understood as tending to unknowingly make ill-grounded or unwarranted claims, some recent epistemological theories in philosophy are relevant because they allow us to examine what it means to have an ill-grounded belief. Memory and knowledge are intimately connected; some portion of our knowledge is embodied in our memory. Psychologists study knowledge, the gaining and forgetting of it, the storing of it in memory, as well as its role in thought, so I will make use of their findings also. Philosophers, on the other hand, have focused on questions such as, How certain do knowledge claims have to be? and What criteria must be met before someone can be said to know something? More specific philosophical questions also arise as to how reliable people's reports are of what they believe or perceive. Confabulation casts doubt on certain types of reports, including reports of memories and the condition of the body.

We need to build several bridges of thought between philosophy and neuroscience to let them illuminate each other. As I noted earlier, it was not always this way, and encouraging signs suggest that it will not continue to be so. The argument is often given that the different disciplines had to be created because of the amount of information that was accumulating. It simply became impossible to know about all of it; there was more information than one person could grasp. Perhaps for this reason, and with the ascendance of behaviorism in the 1930s, philosophy moved, with psychology, away from neuroscience.

In the 1950s, however, breakthroughs in computer science allowed for the first time the creation of an effective model of what was happening inside the behaviorist's black box. This led to the rise of artificial intelligence in the 1960s and 1970s. In philosophy, beginning in the 1980s, Paul and Patricia Churchland began pointing out to their colleagues that information coming out of the neurosciences was increasingly relevant to age-old philosophical problems about consciousness and perception, about responsibility, and self, as well as to many other problems involving the mind. They created a subdiscipline within philosophy, *neurophilosophy*, in order to study these connections (see P. S. Churchland 1986; Hirstein 2004).

Philosophy does move forward, but more slowly than science, and often more because of slow paradigm changes than the clear breakthroughs seen in science; people slowly move to the view they find more interesting. Science always retains the prospect of quick and decisive refutation, something rare in philosophy, barring an obvious mistake in logic. The exact nature of the relation between philosophy and science is a contentious issue among philosophers. There are two competing conceptions of what philosophy itself is, each with a different notion of how close philosophy is to science. Our subject matter spans the two disciplines, so it is important to be clear on this issue. According to the first conception, the distinction between philosophy and science is clear cut; according to the other, the two merge gradually into each other at their boundaries. According to what we might call the "armchair" conception, philosophy is concerned with a priori truth, truth discoverable without experience of the world. Thus, philosophical problems can be solved from an armchair using thought alone, aided by thought experiments in which certain possibilities are imagined and conclusions drawn from them.

According to the opposing concept, philosophy and science are merely different phases of a question-answering, knowledge-gathering process. Philosophers deal with questions at their early, murky stages, often working closely with scientists, toward the goal of handing over a clearly defined, empirically researchable problem to science. As Paul Churchland put it, "the philosopher is just another theorist, one whose bailiwick often places him or her at the earliest stages of the process by which proto-scientific speculation slowly develops into testable empirical theory" (P. M. Churchland 1986, 6). I will act as such a theorist here, debating issues with scientists when I feel qualified to, or stepping back and focusing instead on arranging their voices into a perspicuous presentation of the topic.

Philosophy's job is sometimes just to make the questions themselves clear, and in so doing to develop techniques for posing the questions and their answers. This second conception of philosophy, in which it is continuous with science, has two things going for it. First, it makes better sense of the history of philosophy. In the armchair view, much of that history seems to be miswritten, since the first philosophers addressed questions that are today regarded as scientific such as, What is the nature of matter?, or What causes weather? Second, the idea that using one's mind alone to solve a problem does not involve observation of the world seems to assume mind-body dualism. Contrary to this, materialists believe that the mind is also part of the world, so that observing the mind, as one does in armchair introspection, counts as observing the world.

Materialism is a philosophical theory that, when applied to the mind, holds that the mind is entirely physical (Armstrong 1968). According to the materialist, all of the amazing qualities of our mental lives are actually physical properties. According to this view, we can learn a great deal about the phenomena we associate with the mind from scientific theories in disciplines such as biology, chemistry, and physics. Materialism is the philosophical position most consonant with these sciences. It merges well with neuroscience, and it allows the formation of a consistent view about the mind across philosophy, psychology, and neuroscience. Most psychologists and neuroscientists are, understandably, materialists. Materialists may be less common among philosophers, but they seem to predominate there also, albeit in a bewildering variety of competing types. Our awareness of our minds, typically said to be accomplished by a mysterious process known as introspection, must also have a physical basis for materialists. They differ greatly, however, in their views of how reliable reports based on introspection are.

Some materialist theories of the mind are consistent with the ways in which we currently think of our minds, but other theories imply that these folk ways of thinking are radically mistaken. The other view the Churchlands are famous for is eliminative materialism (P. M. Churchland 1979). They proposed that in the future we will not speak of beliefs or desires, but rather will adopt entirely new ways of speaking about the mind based on the findings of neuroscience. At this point it seems better to be circumspect and hold to the claim that certain properties of the mind can be given a satisfying physical-biological explanation. Whether other approaches directed at different sets of properties—literary, artistic, historical, political, and social—might also have something to offer in the attempt to explain the mind is thus left completely open. Ideally, one would string together the best theories from each approach in a coherent way, something eliminativists call the Panglossian prospect (Stich 1983), after Voltaire's character Dr. Pangloss, who insisted that ours is the best of all possible worlds.

There is potentially great benefit for philosophers in these borrowings from neuroscience, in the form of information relevant to their age-old problems, some of which may now be riddles. Typically, we already possess the information needed to solve riddles. In Sophocles' play *Oedipus Rex*, for example, when Oedipus is asked by the Sphinx, "What walks on four legs in the morning, two legs at noon, and three legs at night?" he already knows that babies crawl, adults walk on two legs, and old people sometimes require a cane. He merely has to realize that these are the crucial pieces of information. Riddles require not more data but a new way of looking at the existing data. Without conducting a full analysis, many problems having to do with the mind seem to satisfy this criterion. If the problem of explaining confabulation is a riddle, one suspects that it has not been one for very long. What was needed was more accurate knowledge about where the brains of confabulators are damaged. At the same time, effective theories of the functions of these areas had to be developed.

2.5   Approach of This Book

Despite their surface differences, both science and philosophy employ something like the scientific method. Hypotheses are developed, tested, and modified or rejected, and around it goes. Specific techniques for testing differ, of course; scientists typically need a laboratory and expensive equipment, while philosophers usually use their minds as a lab. Their equipment consists of both natural facility at reasoning and special philosophical techniques they have learned, for analyzing arguments, concepts, the semantic properties of language, and so on. Scientists construct models and simulations of phenomena of interest, but philosophers have been doing this for millennia. In a philosophical thought experiment, a certain situation is described, and the philosopher uses the appropriate techniques to discern its implications (typically logical or metaphysical implications).

The Churchlands believed that a more effective theory of the mental should be modeled on the way groups of neurons actually behave; that is, it should be a neuroscientific theory. An approach based on available information about confabulation, however, employs a different and higher level of explanation in neuroscience. Whereas the Churchlands focused their attention primarily on the field of neurocomputation, I will focus on fields a level or two "up" from there: cognitive neuroscience, cognitive neuropsychology, and gross neuroanatomy.

I will employ two basic techniques for explicating connections between neuroscience and philosophy. The first is to connect some phenomenon or hypothesis in neuroscience with a phenomenon or hypothesis in philosophy through a chain of inferences. The second technique involves carefully arranging the data and hypotheses from the two fields in such a way that their connections can be easily seen, assembling different data into a perspicuous representation of the facts involved. The order of exposition will tend to be from the concrete to the abstract, in the sense that I will begin with scientific facts and move to philosophical arguments, especially when I construct definitions of confabulation and examine their relation to existing definitions of self-deception. Many other philosophical problems arise along the way that space considerations prevent me from delving into: problems involved in the concept of memory, in the notion of a body image, in our knowledge of ourselves and our consciousness. For the time being, however, we must focus on data collection and organization in order to understand the basic phenomena of confabulation.

Chapters 3 through 7 are devoted to describing and analyzing what is known in neuroscience about different confabulation syndromes. These chapters are primarily data driven, but I have attempted to arrange and connect the data so that they form a coherent picture. In chapter 3 I suggest that one can find in the data a rough consensus view about the nature of neurological damage in memory confabulation syndromes. In chapters 4 through 7, I endeavor to show that this consensus view can either account for the other confabulation syndromes or can be augmented to do so. Chapter 4 is an examination in greater detail of the functions of a brain area that comes up frequently in chapter 3 when crucial lesions for confabulation are discussed: the orbitofrontal cortex. I propose that a comparison of confabulations with the communications of sociopaths, who have been found to show compromised orbitofrontal function, and also with lying by normal people, helps illuminate the nature of confabulation.

Chapter 5 is devoted to confabulation as it occurs in misidentification syndromes, such as Capgras' syndrome. There I suggest that the way to connect these cases with other cases of confabulation is to see them all as involving theory of mind or mind-reading deficits—different deficits in the brain's systems devoted to modeling the thoughts, intentions, and actions of other people—and that this explains why confabulators are oblivious to the incredulity of their listeners. The denial of illness that occurs when people falsely claim they are not paralyzed or blind is examined in chapter 6. Chapter 7 looks at confabulation by split-brain patients, and at the theories of Geschwind and Gazzaniga about why it happens. In chapter 8 I distill all of these findings into a theory of confabulation that views it primarily as an epistemic phenomenon. A person is confabulatory when he or she has a tendency to unknowingly make ill-grounded claims that he or she should know are ill-grounded. Typically this tendency is confined to a particular knowledge domain, such as memory, knowledge of other people, or knowledge of the body.

Once this theory of confabulation is in place, I apply the findings from chapters 3 through 7 as well as the theory in chapter 8 to the problem of self-deception. There I argue that once confabulation is seen as a kind of disinhibition, we can see the nature of the two forces at work in self-deception. The creative process that is disinhibited and unopposed in clinical confabulators is opposed in self-deceived normal people, but the processes capable of dislodging the self-deceptive beliefs are temporarily held at bay by intentional or nonintentional means.

3 Confabulation and Memory

When asked to tell how he has been spending his time, the patient would very frequently relate a story altogether different from that which actually occurred, for example, he would tell that yesterday he took a ride into town, whereas in fact he has been in bed for two months, or he would tell of conversations which never occurred, and so forth.
—Sergei Korsakoff (1889)

3.1 Fictional Autobiographies

What did you do yesterday? To answer this question, you employ your *autobiographical memory*. Our autobiographical memory is our record of our daily experience. It is highly biased in that its most detailed representations are of the events most significant to us. Autobiographical memory is also part of our sense of identity as beings with a determinate past, stretching continuously back in time. It is also our catalog of specific episodes in which we interacted with significant people. This type of memory is the one most commonly compromised in amnesia, such as in patients with Korsakoff's syndrome, who have no memory at all of what they did yesterday. We have all experienced the memory malfunctions one sees in different types of amnesia. We sometimes remember what we intended to say or do, rather than what we actually said or did. We frequently displace events in time on recalling them. And then there is the interesting error of mistaking events that were merely dreamed about for real events, and the perhaps rarer case of mistaking a memory of real events for a memory of dream events. Trends in memory research strongly confirm what memory researchers have always known, that the distinction between normal, accurate memory reports and memory confabulations is not that simple. Remembering is not merely the transmission of the appropriate information from storage into consciousness. Rather, it is a reconstructive process (Schacter et al. 1998; Loftus and Pickrell 1995) that can go wrong in any of several ways.

Confabulation caused by frontal brain injury is the best-studied type. The findings suggest that some such damage is essential for the phenomenon, so this is the best place to start our review of the syndromes. This damage produces confabulations about past events in the patient's life that did not happen, did not happen to him, or did not happen to him when he believes they did. This chapter gives a detailed analysis of confabulation syndromes that involve serious memory deficits, including Korsakoff's amnesia, and aneurysm of the anterior communicating artery.

Since the mid-1900s we have learned a great deal about the brain areas and functions that comprise autobiographical memory. It was discovered by accident that certain parts of the temporal lobes play a crucial role in what we might call indexing, or managing this memory. Memory deficits caused by aneurysms of the anterior communicating artery and by lifelong drinking in patients with Korsakoff's syndrome hold a special interest for researchers, because such deficits indicate that this memory system also has important frontal components. Sites of lesions in Korsakoff's and anterior communicating artery syndromes are clearly different from those involved in medial temporal lobe amnesia. Anatomists confirm the idea that the frontal and temporal lobes work together to achieve memory, by finding that areas that constitute the medial temporal lobe memory system have strong reciprocal connections to at least two frontal areas. Patients with temporal and frontal memory disorders differ in important ways, however. Those with medial temporal lobe disturbances do not confabulate (Parkin 1984; Moscovitch and Melo 1997). In fact, they were found to be *less* likely than normal people to produce false memories on a task specifically designed to elicit them (Schacter et al. 1996b). Patients with medial temporal lobe amnesia also showed much higher latencies in giving answers and made many more self-corrections than amnesia patients with frontal damage in a memory task (Mercer et al. 1977).

With increasing information available about how these memory systems work, the discussion of memory-based confabulation has grown increasingly sophisticated. One theme of great interest that comes up frequently in the literature is the idea that these types of confabulations might be caused by two separate malfunctions. First, patients have a memory problem, which they share with those with medial temporal lobe damage. But second, they have what is typically referred to as an executive problem, which is responsible for the failure to realize that the memories they are reporting are fictitious. In a particular case, the two problems manifest as two phases. First a false memory is produced, but then frontal areas fail to perform functions that would allow the person to realize the falsity of the memory. Theories of the nature of the problem in memory confabulation divide into two categories, depending on which of the problems they emphasize:

1. Retrieval theories   Confabulation is caused by a deficit in the "strategic retrieval" of memories (Moscovitch 1995). This causes loss of the sense of the temporal order of one's memories and their sources—the places and times they represent. These theories can be traced back to Korsakoff himself (1889).

2. Executive theories   These theories typically acknowledge that amnesia is present, but add that confabulators are to be differentiated by additional frontal damage (Stuss et al. 1978; Fischer et al. 1995; Kapur and Coughlan 1996b; Burgess and Shallice 1996). According to these theories, two differ-

ent processes are damaged in confabulation: a memory process and an executive or "monitoring" process.

Cognition requires both representations and processes for manipulating those representations. These are executive processes and they perform many different operations on representations. Our memory itself is just a huge collection of representations; executive processes must control the search and reconstruction process that takes place when we remember. Marcia Johnson's view, according to which confabulation is attributed to a deficit in a more general executive function she calls "reality monitoring," (the ability to distinguish real from imagined events) is an example of an executive theory (Johnson 1991). Normal people are able to differentiate real from spurious information at high success rates. This seems to be a learned, or at least a developed, ability. Small children often have trouble drawing a line between the real and the imagined.

The search for the neural locus and functional nature of these executive processes begins with what is known about executive processes that are affected when the lateral portions of the prefrontal cortex are damaged. One line of inquiry involves determining how often confabulation tends to occur in the presence or absence of other mental functions known to require frontal activity. In the Wisconsin Card-Sorting Test, the patient must first sort cards by the color shown on them; then the rule is changed so that the cards are to be sorted by the shape of the figures on them, and so on. Patients with lateral frontal damage get "stuck" responding the first way, and are unable to change to the new sorting rule, a phenomenon known as perseveration. However, several confabulating patients have been described who perform normally on standard tests of frontal function such as this one, causing speculation that confabulation and failure on these tests are the result of damage to different frontal areas.

The selectivity of confabulation becomes a vital issue in the discussion of frontal theories; that is, what does the patient confabulate about, and what does he not confabulate about? For instance, a basic question concerning confabulation about memories is whether it occurs with all of the different memory systems. Does it encompass not only episodic-autobiographical memories about events in one's personal history but also *semantic memories* about impersonal facts, such as knowledge of who the current president is? Whereas the deficits in Korsakoff's syndrome and anterior communicating artery aneurysms are traditionally found in autobiographical memory, several clearly documented cases described patients who also confabulated in response to questions tapping semantic memory. In their review of confabulation syndromes, Johnson et al. stated that "confabulations can be about autobiographical events, beliefs, or semantic knowledge; they can refer to the more remote past, the more recent past, the present, or the future" (2000, 368). Given that confabulation may involve two different types of damage, the problem of selectivity is doubled. There is selectivity in the amnesia, since not all memory systems are affected, but there is also the question of selectivity at the executive level.

3.2 The Brain's Memory Systems

Ever since the pioneering work of French scientist Theodule Ribot, neuroscientists have known that there is more than one type of memory. Ribot wrote that his patients had severe memory loss for factual information but that "neither the habits, nor skill in any handiwork, such as sewing or embroidery ... disappears" (1881, 473). He explicitly based his approach on the work of Jackson: "Hughlings Jackson," he said, "was the first to prove in detail that the higher, complex, voluntary functions of the nervous system disappear first, and that the lower, simple, general and automatic functions disappear latest. . . . What is new dies out earlier than what is old" (1881, 482).

The recent history of memory research begins in the 1950s with the study of H.M., a man who developed severe amnesia after bilateral surgical removal of the medial temporal lobes (including most of the hippocampus [see figure 3.1], parahippocampal gyrus, and amygdala), in an effort to lessen the severity of his epilepsy (Scoville and Milner 1957). H.M. retained his basic intelligence and personality but lost the ability to remember anything that happened to him after the operation. Researchers noticed, however, that he could retain information for a short time, and could also acquire new motor skills such as mirror writing, solving puzzles, and tracing mazes, without knowing that he was doing so. Cohen and Squire (1980) called this form of memory for skills *procedural memory*. This discovery reinforced Ribot's original idea that the human brain contains many memory systems.

The next important step in uncovering the neural bases of these memory systems occurred when the first animal model of human amnesia was established in the rhesus monkey, the highest-level primate on which invasive brain experiments may be performed. Once an animal model exists, researchers can work step-by-step, deactivating (by cooling or electrical impulses) or destroying different areas, and observing the effects of this on the animal's memory. As this process unfolded, it became clear that the medial temporal lobes form part of an extensive memory system.

We have learned a huge amount from animal models, but how do you find out how good a rhesus monkey's memory is? The most popular way to do this is called the delayed-nonmatching-to-sample task. First a sample object is presented, a red cube for instance. Then, after a delay, the monkey is shown two objects, the original one and a new one, a green ball. It must displace the unfamiliar object, the nonmatching object, to receive

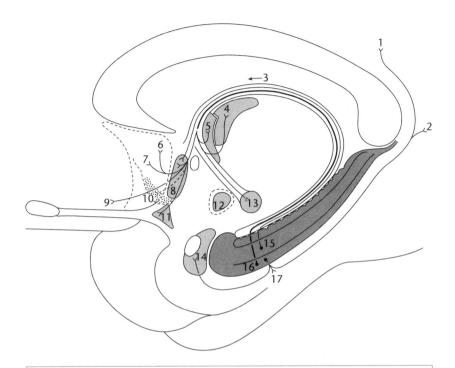

1 Gyrus cinguli (area 23)
2 Cortex retrosplenialis (areae 29, 30)
3 Fornix
4 Nucleus anterior thalami
5 Nucleus interstitialis striae terminalis
6 Hippocampus precommissuralis
7 Cortex frontalis medialis
8 Nuclei septi
9 Gyrus rectus
10 Nucleus accumbens
11 Nucleus olfactorius anterior
12 Nucleus ventromedialis
13 Corpus mamillare
14 Corpus amygdaloideum
15 Cornu ammonis
16 Subiculum
17 Cortex entorhinalis (area 28)

Figure 3.1
Hippocampal connections to brain areas damaged in Korsakoff's syndrome.
Areas of interest here are (3) fornix, (4) anterior thalamic nucleus, (8) septum,
(9) gyrus rectus, part of the orbitofrontal lobes, and (13) mamillary bodies. Notice how the fornix divides into an anterior portion, known as the column, and
a posterior portion that connects to the mamillary bodies. (From Nieuwenhuys
et al., 1988, with permission.)

a food reward, usually fruit or juice. The first approach was to produce a lesion in the monkey similar to the one that H.M. had. Sure enough, this lesion, which damaged the hippocampus, amygdala, and underlying cortex, produced a dense amnesia. The amnesia was multimodal, just as with humans, and as with humans, it did not affect the monkey's short-term memory or memory for certain skills. Subsequent studies in which smaller areas were damaged appeared to show that damaging the amygdala had little effect on memory, whereas the amount of the hippocampus damaged correlated well with the severity of memory loss, especially when the damage was extended forward to include the perirhinal cortex.

Psychologists make a division between two types of memory in the brain. With some things you know, the knowledge of when and where you first acquired that information is long gone. You know that cats have claws, but you most likely have no idea when you learned this. Other information brings with it what researchers call a source memory: memory about when and where that memory was first acquired. Source memory is a fragile thing, and we are prone to characteristic errors in source memory tasks. Schacter et al. (1984) found that patients with frontal lobe damage did poorly on a task in which they had to listen to trials in which one of two people stated a fact, then were later asked to retrieve the information and indicate which of the two people stated it. Shimamura et al. (1990) had normal people and those with frontal lobe damage learn the answers to a set of trivia questions. When they were tested a week later, the latter had normal memory of the answers but showed poor source memory, often claiming they had learned the answer at an earlier time in life.

These are considered episodic memories; semantic memories, such as that the Eiffel Tower is in Paris, or that Harry Truman was a U.S. president, do not normally include source tags. These facts are represented from an impersonal perspective, whereas autobiographical memory is essentially personal. Together with episodic memory, semantic memory forms your *declarative* (or explicit) memory. This distinguishes it from the nondeclarative (or implicit) memory systems, such as those that support learning a skill. So far, researchers have been unable to separate the neural loci of semantic and episodic memory, and perhaps for good reason. The two memory systems interact in several ways, and perhaps the two "can be conceived as different levels of categorization in one and the same memory store" (Fuster 1995, 17).

The medial temporal lobe memory system is thought to include the hippocampal formation (which consists of the CA fields of the hippocampus, dentate gyrus, subiculum, and entorhinal cortex; figure 3.1, numbers 16 and 17) and the adjacent parahippocampal and perirhinal cortices. The hippocampus is not the place where the content itself of memories is stored, but rather appears to contain a set of neural links to the content, which is distributed widely throughout the cortex. Memories of an episode in one's life typically contain information from more than one modality: vision, hearing, taste, touch, and smell. Each of these components is stored in a unimodal sensory area; for example, the visual components of an episodic memory are stored in the visual cortex in the occipital lobe, and the auditory components are stored in the auditory cortex in the temporal lobe. These distributed representations are linked to an index in region CA3 of the hippocampus (Treves and Rolls 1994; McClelland et al. 1995). When recent episodes are retrieved, cues reactivate the index, causing activation to spread to each of the associated unimodal areas. This is more correct of recent episodes, however. Once a representation of an episode has been fully "consolidated," activation can spread among the separate features so that the hippocampus is no longer needed.

We are also beginning to understand the structure, connectivity, and function of the brain areas that make up the frontal components of the medial temporal lobe memory system. Kopelman and Stanhope (1997) made an informative comparison between three types of amnesia: temporal lobe, Korsakoff's syndrome, and frontal lobe. They found that the medial temporal-hippocampal regions were strongly involved in spatial context memory, and the frontocortical region, the diencephalon, and (probably) the temporal lobes are involved in temporal context memory.

Much has also been learned about the neural bases of short-term memory systems located in the frontal lobes. Psychologists have had trouble determining whether there is one type of short-term memory or several. The time span involved—exactly what "short" means—is also not widely agreed upon. In the 1980s, however, neuroscientist Patricia Goldman-Rakic (1987) began exploring a large area in the lateral portion of the prefrontal lobes. This area seems to be responsible for what is called working memory.

3.3   Korsakoff's Syndrome

In 1881 German neurologist Carl Wernicke wrote about a form of brain damage in two alcoholic men and a twenty-year-old woman who had attempted suicide by swallowing sulfuric acid (Finger 1994). He made the following observations: these patients are in an acute confusional state; they have ocular disturbances; and they have trouble walking. The confusional state eventually clears, but Wernicke's patients have other problems in the long term, as Sergei Korsakoff noted. Korsakoff was working at approximately the same time, and in 1887 reported his findings concerning a group of patients with polyneuritis, i.e., paralysis of certain muscles, muscle contractions or atrophies, and weakness or staggering. Korsakoff also observed mental problems that included memory loss, anxiety, fear, depression, irritability, and confabulation. In the early 1900s

Bonhoeffer noticed that most of his patients with Wernicke's syndrome later developed Korsakoff's syndrome, and today it is widely thought that Wernicke and Korsakoff were looking at two different stages in the same disorder, now sometimes called Wernicke–Korsakoff's syndrome. Disagreement about whether the polyneuritis and the mental problems always co-occurred led Jolly (1897) to coin the terms *Korsakoff's psychosis* or *Korsakoff's syndrome* to refer solely to the memory problems and the confabulation.

We have since learned that Korsakoff's syndrome is caused by a lack of vitamin B_1, or thiamine, and not directly by alcohol. The syndrome occurs primarily in alcoholics, but may also occur in other patients whose digestive system fails to absorb B_1, in conditions known as malabsorption syndrome and regional enteritis, as well as with cancer of the stomach (Parkin and Leng 1993). The syndrome has also been reported in people engaging in hunger strikes, in anorectics, and in people receiving prolonged intravenous feeding. The exact means by which alcoholism causes Korsakoff's is not fully understood, but alcohol is known to interfere with transport of thiamine in the gastrointestinal tract, and chronic liver disease can also affect the liver's ability to store thiamine (Parkin and Leng 1993). In their authoritative work on Korsakoff's and other memory syndromes, Parkin and Leng noted that thiamine is important for proper brain metabolism, because chemicals derived from thiamine play a role in "the function of excitable membranes, several biosynthetic reactions involved in glucose metabolism, and the synthesis of neurotransmitters, particularly acetylcholine, and [GABA]" (1993, 36). Korsakoff's syndrome can come on quickly, after an alcoholic coma (Torvik et al. 1982), or it can progress slowly over many years (Cutting 1978). Immediate administration of high doses of thiamine can somewhat alleviate the symptoms once the disease is diagnosed.

The memory loss in Korsakoff's is anterograde—the patients are unable to form new memories. Procedural memory is intact; for instance, the patients can still drive. These patients tend to underestimate the time they have spent in the hospital, as well as, remarkably enough, their own age (Zangwill 1953). Korsakoff (1892) successfully traced the memory reports of his patients to actual memories that they had displaced in time. In his classic book *Deranged Memory*, George Talland constructed an interesting list of characteristics of the confabulations of these patients: "In the early phase of the Wernicke–Korsakoff syndrome confabulation is: (a) typically, but not exclusively, an account, more or less coherent and internally consistent, concerning the patient. (b) This account is false in the context named and often false in details within its own context. (c) Its content is drawn fully or principally from the patient's recollection of his actual experiences, including his thoughts in the past" (1965, 49–50).

Many Korsakoff's patients are not well oriented as to where they are or what the time is, and have an odd lack of interest in their surroundings (Talland 1961). They are often unaware that they are in a hospital, for instance. Talland emphasized the role of this disorientation in producing confabulation and gave a good description of it: "In the early stages of the amnesic syndrome the patient is severely confused and disoriented. He does not know or only gradually realizes that he is in a hospital and he has no recollection of the events that resulted in his transfer to the ward" (1965, 45). Many patients briefly accept the idea that they are in a hospital, but then soon revert to claiming otherwise: "Certitude in identifying a place bears no relation to its accuracy. The notion that it is a hospital can be entertained very tentatively or fleetingly; the delusion that it is some rooming house or 'The Eagle which used to be Mrs. Porter's tavern,' can be held firmly day after day, or discarded at the first intimation of doubt" (Talland 1965, 46). Orientation for place tends to improve after the acute phase, but the temporal disorientation tends to persist into the chronic phase.

The complications in Korsakoff's are serious. They include permanent loss of memory and cognitive skills, injury caused by falls, an alcohol withdrawal state, permanent alcoholic neuropathy, and a shortened life span. It is interesting that the patients also have difficulty with social interactions, a symptom they share with many other types of confabulating patients. Parkin and Leng (1993) noted decreased noradrenergic activity, a symptom that other confabulators may also share. There have also been reports of perceptual deficits; one group of patients did poorly on an embedded figures test and had difficulty alternating between the two ways of seeing the Necker cube (Talland 1965). Another characteristic of patients with Korsakoff's syndrome is suggestibility (Berlyne 1972). As an informal test of this, Talland tried asking twenty patients, "Didn't we meet over a drink last night?" The patients were not having any of it. As Talland told it, "although I put it as jovially and convincingly as I could, all the responses were denials" (1965, 56). Mercer et al. (1977) tried a more subtle type of suggestion. After asking confabulatory patients forty-one factual questions, they told the patients that they had given the correct answer to three questions, but that their answers had inadvertently not been recorded. The questions chosen were actually the first three to which the patient had answered, "I don't know." Approximately thirty percent of the patients then changed their original "don't know" response and confabulated answers.

The Neuroscience of Korsakoff's

There is wide agreement on two sites of damage in this syndrome—the thalamic mediodorsal nuclei and mamillary bodies; the importance of other damage sites is more controversial. Some researchers believe that the possibility remains that damage to one of these alone is the minimal lesion for Korsakoff's, that is, it is capable of causing the syndrome by itself. Early writers favored damage to the mamillary bodies as the crucial lesion, but a more recent trend points to the mediodorsal nucleus as the culprit. Parkin and Leng (1993) suggested that the minimal lesion must involve damage to both the mamillary bodies and the mediodorsal nucleus. The most frequently detected lesions in Korsakoff's patients are in the following brain areas: the thalamic mediodorsal nucleus, mamillary bodies, frontal lobes, and anterior thalamic nuclei.

Thalamic Mediodorsal Nucleus   When they examined the brains of patients with Korsakoff's syndrome postmortem, Victor et al. (1971) found cell loss in the mediodorsal nucleus of the thalamus (figure 3.2, number 7, and figure 3.3). This nucleus, sometimes also called the dorsomedial nucleus or medial nucleus, is the area responsible for connecting large areas of the prefrontal cortex, including orbitofrontal lobes, to the thalamus. Some authors even define prefrontal cortex as the area of cortex connected to the mediodorsal nucleus. Different divisions of the mediodorsal nucleus interact with different parts of the prefrontal lobes, however. One conception of the function of the mediodorsal nucleus is that it transmits information from limbic areas to the prefrontal cortex. Gloor posited that the magnocellular division of the mediodorsal nucleus "receives highly processed inputs from mesial limbic structures and from the rostral temporal isocortex, and after further processing relays them to the orbital and mesial frontal cortex" (1997, 253).

The mediodorsal nucleus participates in an important circuit involving the amygdala and orbitofrontal cortex (figure 3.3). Several authors speculated that damage to this circuit is crucial for the production of the executive deficits found in confabulation, but it also apparently has purely mnemonic functions. Destroying any of these three organs in rhesus monkeys had roughly the same negative effect on a reward memory task (Gaffan and Murray 1990). The authors proposed that this supports the claim that the three are functionally related to each other. Miller et al. suggest that the right mediodorsal nucleus is responsible for "inducing and coordinating the recall" of "world knowledge pertaining to people and events" as well as autobiographical information (2001, 1037). Their patient initially had a small, bilateral thalamic lesion that, as it extended into the right thalamus, caused inability to remember autobiographical events more than a few years old, in addition to inability to recognize famous people or important world events. However, this patient did not confabulate in memory tests, suggesting that right mediodorsal damage alone is not sufficient to cause confabulation.

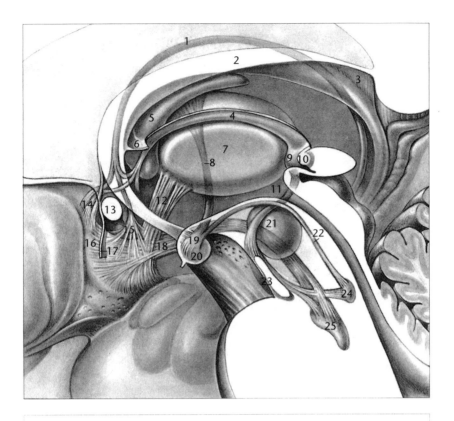

1 Stria terminalis	14 Precommissural ⎰ stria terminalis
2 Fornix	components of ⎱ stria medullaris thalami
3 Commissura fornicis	fornix
4 Stria medullaris thalami	15 Stria terminalis postcommissuralis
5 Nucleus anterior thalami	16 Septum verum
6 Tela choroidea ventriculi tertii	17 Lamina terminalis
7 Nucleus medialis thalami	18 Fasciculus telencephalicus medialis
8 Tractus mamillothalamicus	19 Fasciculus mamillaris princeps
9 Nuclei habenulae	20 Corpus mamillare
10 Commissura habenulae	21 Nucleus ruber
11 Tractus habenulointerpeduncularis	22 Tractus mamillotegmentalis
12 Pedunculus thalmi inferior	23 Nucleus interpeduncularis
13 Commissura anterior	24 Nucleus tegmentalis dorsalis
	25 Nucleus centralis superior

Figure 3.2
Subcortical brain areas of interest in the study of Korsakoff's syndrome and
ACoA syndrome. (From Nieuwenhuys et al., 1988, with permission.)

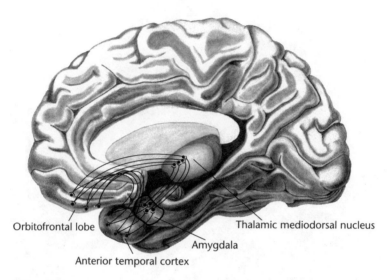

Orbitofrontal lobe

Anterior temporal cortex

Amygdala

Thalamic mediodorsal nucleus

Figure 3.3
Circuit composed of orbitofrontal cortex, thalamic mediodorsal nucleus, and
amygdala-temporal pole. (Based on K. E. Livingston and A. Escobar, 1973. Ten-
tative limbic system models for certain patterns of psychiatric disorders, in *Surgical Approaches in Psychiatry*, eds. L. V. Laitinen and K. E. Livingston, Baltimore,
Md.: University Park Press.)

Mamillary Bodies The mamillary bodies are connected by the fornix to
the hippocampi (see figure 3.1, number 13 and figure 3.2, number 20).
Sclerosis of the mamillary bodies was the earliest anatomical finding in
patients with Korsakoff's syndrome (Gamper 1928). Brion et al. (1969)
found that disconnection of the mamillary bodies from the fornix produced amnesia. A large study of forty-five Korsakoff's patients by Torvik
(1982) confirmed the claims of damage to both the mamillary bodies and
the mediodorsal nucleus.

Frontal Lobes Jernigan et al. (1991) analyzed magnetic resonance images
of patients with Korsakoff's syndrome and found reduced gray matter vol-
umes in anterior diencephalic structures, the orbitofrontal cortex, and the
medial temporal lobes. Benson et al. (1996) found decreased blood flow
in the bilateral orbitofrontal cortex of a patient in the acute stage of the
syndrome.

Anterior Thalamic Nucleus This nucleus can be seen in figure 3.1, number
4 and figure 3.2, number 5. Cell loss has also been seen here in Korsakoff's
patients (Benson 1994). Alcoholics with the syndrome shared with other

alcoholics loss of neurons in the mamillary bodies and mediodorsal nucleus (Harding et al. 2000). Patients with Korsakoff's syndrome were different from the other alcoholics, though, in that they also had a loss of neurons in the anterior thalamic nuclei.

Mishkin (1978) maintained that two separate memory circuits are damaged in Korsakoff's syndrome. The first we are familiar with; it consists of the mediodorsal nucleus, orbitofrontal cortex, and amygdala. The other involves the mamillary bodies, the anterior thalamic nuclei, and the hippocampus. Sectioning the anterior temporal stem and amygdala, and sectioning the fornix each produced measurable but mild impairment (Gaffan et al. 2001). However, when both operations were performed on monkeys, the animals had severe impairment in a memory test that required that they remember shapes against colored backgrounds. This is consistent with Mishkin's hypothesis. The amygdala section disrupted the amygdala circuit, and the fornix section disrupted the hippocampal circuit.

3.4   Aneurysms of the Anterior Communicating Artery

Aneurysms of the anterior communicating artery—which distributes blood to portions of the ventromedial frontal lobes, including parts of the orbitofrontal lobes and related structures, including the basal forebrain (figure 2.2), fornix (figure 3.1, number 3 and figure 3.2, number 2), septum (figure 3.1, number 8), anterior cingulate gyrus (figure 2.2), and corpus callosum—can also cause confabulation. The ACoA is one of the most common sites of aneurysm in the human brain (Carpenter 1985). Aneurysms occur when the walls of a blood vessel are weakened by an insult, such as infection or degenerative illness. Often the vessel ruptures, causing a hemorrhage. Blood itself is quite toxic to brain tissue, and wide and permanent destruction can result. Sometimes brain tissue is also damaged when a surgeon operates to close off a damaged blood vessel. The ACoA forms the anterior link of the circle of Willis, a ring of arteries that spans the inner portions of the two hemispheres and interconnects the two large anterior cerebral arteries. Although the ACoA is small, it feeds a variety of brain areas and organs, and damage to it may also seriously affect blood flow in one or both anterior cerebral arteries. Fiber tracts of several important brain circuits pass through the ACoA's area of distribution (figure 3.4).

Aneurysms of the ACoA are serious, and neurological literature is filled with dire statements, such as, patients "have historically been observed to suffer from a poor neuropsychological outcome" (Mavaddat et al. 2000, 2109). As with Korsakoff's syndrome, in the acute phase "there is a severe confusional state and attention disturbance, and confabulation—sometimes fantastic—with lack of insight may also be found" (Parkin and Leng 1993, 116). The damage does not seem to affect the sorts of cognitive

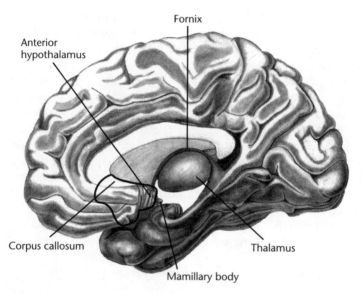

Figure 3.4
Distribution area of the anterior communicating artery. (Based on figure 2 in
R. M. Crowell and R. B. Morawetz 1977. The anterior communicating artery
has significant branches. *Stroke* 8: 272–273.)

abilities the typical IQ test measures, however (Beeckmans et al. 1998).
Testing of their visual abilities has not revealed significant problems, except
in those in whom blood flow to the optic chiasm or optic nerves is disrupted. The important cognitive features of the classic ACoA syndrome are
as follows:

1. Memory loss   Patients show anterograde as well as retrograde amnesia,
often for a few years preceding the aneurysm. As in Korsakoff's syndrome,
short-term memory appears to be intact. In tests of recognition memory,
patients can often correctly recognize famous people at a normal level, but
they can exhibit something called pathological false recognition; that is,
they tend to have a large number of "false alarms" in which they claim to
recognize a stimulus they are actually seeing for the first time.

2. Changes in personality   Just as patients with Korsakoff's syndrome, those
with ACoA disorders have social interaction problems. Impulsivity, impa-
tience, disinhibition, emotional lability, depression, problems in decision
making, and poor judgment in social situations have also been observed
(Mavaddat et al. 2000). These patients may experience loss of self-criticism
and apathy.

3. Executive deficits These include perseveration, poor concept formation, problems with set shifting, reduced verbal fluency, and impairments in cognitive estimation (Beeckmans et al. 1998). Some studies of Korsakoff's patients failed to find a group deficit in the Wisconsin Card-Sorting Test (Mavaddat et al. 2000). These frontal tests employ primarily the dorso-lateral prefrontal cortex, however, and not the ventromedial (or orbito-frontal) areas damaged in ACoA syndrome (Damasio 1996).

4. Confabulation Confabulation occurs in the acute phase right after the aneurysm and often remains in the chronic phase. It can change from im-plausible spontaneous confabulation in the acute phase to more plausible provoked confabulation in the chronic phase.

Just as patients with Korsakoff's syndrome, those with ACoA syndrome become confused about place, time, and even their own identity, especially in the acute phase (Beeckmans et al. 1998). Stuss et al. provide the following example: "Exemplary in this patient's confabulation was a striking misuse of environmental cues.... During one interview, when asked if he knew the place, he responded by stating that it was an air-conditioning plant (he was facing a window air-conditioner). When sur-prise was expressed about his wearing pajamas, he responded, 'I keep them in my car, and will soon change into my work clothes'" (1978, 1167). This disorientation includes topographical disorientation, in which patients are unable to find their way back to their hospital room. Those with ACoA syndrome also tend to have anosognosia (denial of illness) in that they deny there is anything wrong with their memories (Fischer et al. 1995).

Damage to the ACoA is the most focal damage known to cause con-fabulation, so it is a valuable clue to the identity of the crucial brain pro-cesses damaged. Let us look in more detail at what is known about areas likely to be injured in an aneurysm of an anterior communicating artery. Serizawa et al. (1997) divided the areas supplied by the ACoA as follows:

1. The subcallosal area This is the ACoA's largest branch. By way of the median callosal artery, it feeds the rostrum and genu of the corpus cal-losum, the anterior commissure (figure 3.2, number 13) the anterior cingu-late gyrus (figure 2.2), paraolfactory gyrus, and paraterminal gyrus (these last two are visible at the left edge of figure 3.2), the septum, and the col-umn of the fornix. The frontal portion of the fornix, or column, is a possi-ble area of overlap between Korsakoff's and ACoA syndromes. Notice that damage here and to the orbitofrontal cortex disrupts both of Mishkin's memory circuits.

2. The hypothalamic area This includes the anterior hypothalamus (figure 3.4) and the lamina terminalis (see figure 3.2, number 17).

3. The chiasmatic area   This includes the optic chiasm and superior part of the optic nerves.

Parkin and Leng concluded that "the most frequent lesion site involves one or more aspects of the frontal lobes with associated deficits in the basal forebrain" (1993, 116), and several writers argued that amnesia caused by the aneurysm is due to basal forebrain damage (see DeLuca 1993). The phrase "basal forebrain" is used in two different ways, both of which are relevant to confabulation. It can refer to the portion of the cortex just anterior to the hypothalamus, that is, the posterior base of the orbitofrontal lobes, but it also has a more specific reference to the septal nucleus, nucleus accumbens, and nucleus basilis of Meynert, important parts of the brain's cholinergic system (Johnson et al. 2000).

The Neuroscience of ACoA Syndrome
Since the ACoA communicates between the two large anterior cerebral arteries, damage to it can affect blood flow in them. The anterior cerebral arteries or their tributaries distribute blood to large areas of the frontal cortex, including "internal frontal, fronto-polar, and orbitofrontal cortex, paracentral, superior and inferior parietal cortex, the pericallosal region and the basal ganglia" (Parkin and Leng 1993, 115). The anterior commissure and anterior corpus callosum are sometimes also affected by ACoA aneurysm, but I defer discussion of them until chapter 7, since they are also destroyed in certain types of split-brain operation.

Orbitofrontal Cortex   The nearest large artery branching off the anterior cerebral arteries after they are joined by the ACoA is the frontobasal medial artery, which runs directly forward along the orbitofrontal lobes. Disruption of blood flow there can damage the orbital and medial surfaces of the frontal lobes (Parkin and Leng 1993). Bilateral medial frontal destruction was reported in patients with ACoA syndrome (Parkin and Leng 1993). Feinberg and Roane's patient showed "extensive frontal lesions that included the medial, orbital frontal regions of the brain bilaterally" (1997, 81). The orbitofrontal cortex is examined in much greater detail in chapter 4.

Anterior Cingulate   The anterior cingulate (figure 2.2) has functions in attention, emotion, and motor control. It has dense interconnections with the hippocampus, as well as with anterior medial temporal cortices involved in memory (Barbas et al. 2002). Activity was seen in the anterior cingulate during recall of episodic memories (Nyberg 1998), and monkeys with damage there had visual recognition problems (Bachevalier and Mishkin 1986). It is interesting that the anterior cingulate also receives "massive" projections from somatosensory areas (Damasio 1999). One way to gain insight into its role in motivation and motor function comes from noting that bilateral damage can produce a syndrome known as akinetic mutism, in which all motivation seems to disappear. The patient sits calmly and looks at you, but will not respond to any questions or commands. The anterior cingulate is often active during tasks that demand conscious attention (see Posner and Raichle 1994; Devinsky et al. 1995). For instance, increased activity was observed in the anterior cingulate in experimental conditions in which subjects had to choose among competing responses (Carter et al. 1998).

In the 1950s, surgeons began experimenting with lesioning parts of the anterior cingulate gyrus in an attempt to relieve patients with certain problems, typically severe obsessive-compulsive disorder. One strange result of these operations was an inability in some patients to tell the difference between fantasies and reality in the days after surgery. Whitty and Lewin (1957) described this "vivid day-dreaming" after anterior cingulectomy in eight of ten patients. One of these patients, a fifty-eight-year-old businessman, underwent the operation in an attempt to relieve "a severe tension state with obsessional and depressive features for over five years" (1957, 73). The operation produced relief from these symptoms for three years, after which the patient relapsed. He had the following interaction with his doctor forty-eight hours after his operation (1957, 73):

In response to questions as to what he had been doing that day, he replied: "I have been having tea with my wife." Without further comment from the questioner, he continued, "Oh, I haven't really. She's not been here today. But as soon as I close my eyes the scene occurs so vividly, I can see the cups and saucers and hear her pouring out, then just as I am lifting the cup for a drink, I wake up, and there's no one there."
Q.   Do you sleep a lot then?
A.   No I am not asleep—it's a sort of waking dream ... even with my eyes open sometimes. My thoughts seem to be out of control, they go off on their own—so vivid. I am not sure half the time if I just thought it, or it really happened.

A second patient age 31, had a "completely disabling" "severe obsessional neurosis" that centered around a morbid fear that she might swallow pins (Whitty and Lewin 1957, 73). The cingulectomy seemed to help, allowing her to lead "a normal wage-earning life, though some symptoms were still present." In this interview that took place twenty-six hours after surgery, the woman was "slightly uninhibited," and still expressed fear of swallowing pins (1957, 74):

She said: "The nurses drop pins everywhere, in my bed and clothes."
Q.   Do they really?
A.   Well, I think they do.
Q.   Have you seen them doing it?
A.   No exactly; but when I doze off I see it.
Q.   You mean you dream it?
A.   It's more thinking than dreaming; but that's the trouble, my thoughts seem so vivid I can't remember what I think and what happens.
Q.   How did you sleep last night?
A.   All right, I suppose; but there was a lot of noise and two people talking about me just outside the door.
Q.   What were they talking about?
A.   About my symptoms. It was two of the doctors.
Q.   Were they really there?
A.   I seemed to hear them whenever I dropped asleep, but they may have been part of my dreams. I was dreaming all night, and I felt it was all really happening.
Q.   But you could tell if you really heard them.
A.   Yes, I can work it out. But ever since I left the [operating] theater I just can't distinguish between phantasy and reality.

According to Whitty and Lewin, this condition is due to an increase in the vividness of thoughts and dreams, rather than to a decrease in the vividness of perception. They also make the connection that "many normal children under 5 years of age have great difficulty in distinguishing between phantasy and reality" (1957, 75).

Hypothalamus   The ACoA supplies blood to the anterior portion of the hypothalamus, which as we saw in chapter 2 is an important part of both the limbic and autonomic systems. It is also significant that cases of confabulation appear to be caused by damage just to the hypothalamus. An amnesic patient with a medial hypothalamic lesion was disoriented and confabulated (Ptak et al. 2001). Possible roles of the hypothalamus in confabulation are discussed in chapter 4.

3.5   Frontal Theories of Confabulation

Some frontal theories of confabulation can be expanded to fit other confabulation syndromes that do not involve memory problems, but others are so strongly memory based that they cannot be. Theories that attribute confabulation specifically to a memory retrieval problem apply only to memory patients. Korsakoff (1892), for instance, stated that confabulations are simply veridical memories placed out of context. Schnider et al. (1996) concluded similarly that spontaneous confabulations arise from failure to know the proper temporal place of the recalled memory. Other executive theories have the potential to be expanded beyond mnemonic confabulation. In the theory of Marcia Johnson and her group, for instance, confabulation is due to a deficit in a more general executive function they called "reality monitoring." In their view, the distinction between real and imagined extends beyond memory; it can include distinguishing hallucinations from actual perception.

It may be that retrieval theories and executive theories are directed at different parts of the confabulation process. The first phase involves production of a false memory. The second phase involves failure to notice and correct the falsity. Retrieval theories focus on failure to access the correct memory, whereas executive theories focus on the correction failure. Executive theorists typically attribute confabulation to failure in what they call self-monitoring or self-awareness. Mercer et al. (1977) noted, for instance, that all cases of confabulation are based on a failure to self-correct. Benson et al. connected these functional problems to specific brain areas, saying that their research suggests that "confabulation represents a frontal executive control dysfunction and that the cingulate and medial orbital frontal areas are active in the monitoring of verbal responses. When these frontal areas lose basic limbic input, the patient's inability to self-monitor allows confabulation to occur. Confabulation and defective self-awareness apparently represent the same basic functional disorder" (1996, 1243). How exactly do we monitor our beliefs or other representations?

Reality Monitoring: Johnson's Theory

Confabulation may be due to a broader failure to test representations, whether or not they are from memory. The phrase "reality monitoring" seems misleading since what is being monitored (or not) is a representation, often one to which no reality corresponds; "representation monitoring" would be more accurate. Real memories, according to Johnson, can often be distinguished from imaginings by the amount of perceptual detail they contain, as well as by the presence of supporting memories, including information about where and when the remembered event occurred; that is, source memory. Johnson stated that episodic memories of an event bind together elements of several different types, some of which represent properties of the event, whereas others represent features of us; for example, our thoughts or emotions in reaction to witnessing the event. These different properties include colors, sounds, tastes, emotions, objects, and locations, but also information contained in semantic memory. Recall of any one of these features is often enough to draw the entire autobiographical memory back into awareness. When thoughts that have this rich detail present themselves as memories, this can be sufficient to make us regard them as genuine. Because of this, in the mind of a person with a vivid and detailed imagination, memories of imaginings can be mistaken for memories of actual events, rather like the vivid day-dreaming experienced by some patients after anterior cingulectomy.

Johnson et al. say that "a vivid imagination can exceed the threshold and be taken to be a memory of an actual event, for example, when one believes they said or did something that they only thought or imagined saying or doing" (2000, 361). Indeed, Dalla Barba et al. (1990) had a patient attempt to distinguish actual events in her life from confabulations she had produced during earlier testing. She claimed the confabulations described real events eighty-six percent of the time, but accepted the actual events as real only forty-three percent of the time.

Further checks can be made. We can check the consistency of the candidate memory with our set of beliefs. Monitoring can involve noting inconsistencies "among activated information or between activated features and long-term knowledge" (Johnson et al. 2000, 362). DeLuca (2001) described a patient with ACoA syndrome who contradicted himself in the same sentence, saying that he had just visited a store he formerly owned, then acknowledging that the store no longer existed. As early as 1915, Pick noted that patients with Korsakoff's syndrome also have a deficiency in the need to correct contradictions. Johnson et al. note that "events that fit with what we know or believe are easier to encode, easier to revive as memories, and less likely to be questioned later" (2000, 361). However, Mather et al. (1999) concluded that the less actual memory information available, the greater the influence of stereotypes and other beliefs, indicating that in some cases the checking procedures themselves may cause inaccuracies.

People can intentionally tighten their monitoring standards when motivated to do so. "[T]he fact that people sometimes normally adopt less than optimal criteria ... is demonstrated when instructions or test conditions induce them to consult better or additional information and their likelihood of false memories decreases" (Johnson et al. 2000, 362). It is interesting to note that we tend not to loosen our standards voluntarily; rather, it tends to happen unconsciously and spontaneously. Johnson and her colleagues distinguished between *heuristic* checking of candidate memories, which usually operates when we are remembering, and *systematic* checking, which is "engaged selectively and deliberately." "Heuristic processing is comprised of fewer component processes and uses readily available information (e.g., familiarity), including qualities (e.g., perceptual detail) and schemas (e.g., world knowledge, stereotypes) activated by a cue" (Johnson et al. 2000, 362). Systematic processing "tends to be comprised of more component processes" and may also involve the retrieval of "other memories and knowledge that are not readily activated by the initial cue" as part of the checking process (2000, 362).

Systematic processing requires selective attention; the person must explicitly attend to the candidate memory. It also includes self-provided memory cues. We often cue our own memories; when I want to remember someone's name, I imagine her face. I produce a cue for my memory system to use in retrieving the name. I then monitor any representations that the cue gives rise to. I may have to use other information to reject candidate names that come up. Often this cuing process must be used several times in order to reconstruct the memory correctly. "Recall of complex autobiographical memories often involves maintaining an agenda and iterations of a number of these various process—refreshing activated information, initiating cues, retrieving additional information, noting whether its qualitative characteristics meet expectations, discovering consistencies or inconsistencies with other activated knowledge and beliefs and so forth" (Johnson et al. 2000, 363). As to the neural locus of these monitoring processes, these authors point to bifrontal areas: "Self-monitoring is subserved by frontally mediated processes ... perhaps typically involving both hemispheres" (2000, 374).

Suppression of Irrelevant Memories: Schnider's Theory

Talland wrote that "in the Korsakoff syndrome ... frames of reference no longer appropriate are insufficiently inhibited" (1961, 371). Armin Schnider's research group similarly hypothesizes that the problem in memory confabulation is that the orbitofrontal cortex and its limbic connections are not performing their function of suppressing or inhibiting recalled memories that are not relevant to the current task. The posterior medial orbitofrontal cortex "sorts out the mental associations that pertain to ongoing reality by suppressing memory traces that have no current relevance" (Schnider et al. 2000, 5884). As to the question of what the crucial lesion is for producing confabulation, Schnider puts his money on the orbitofrontal-mediodorsal-amygdala circuit (figure 3.3): "Spontaneous confabulation appears to emanate from interruption of the loop connecting the posterior orbitofrontal cortex directly (via ventral amygdalafugal pathways) and indirectly (via the mediodorsal thalamus) with the amygdala" (2001, 155). The group's discovery of a confabulatory patient with damage limited to the hypothalamus (see Ptak et al. 2001) makes Schnider include this part of the brain under the cloud of suspicion: "In addition, the medial, probably anterior, hypothalamus, which has direct connections with the posterior medial OFC [orbitofrontal cortex], also appears to be important" (2001, 155). Schnider's way of connecting confabulation in Korsakoff's syndrome with that found in ACoA syndrome is to point out that the basal forebrain lesions one sees in ACoA aneurysms "often include the posterior medial OFC" (Schnider 2001, 154).

Schnider's localization is supported by two findings. First, "patients with lesions involving the posterior medial OFC and base forebrain confabulated for much longer periods (several months) than patients with anterior medial OFC lesions" (Schnider 2001, 156). And second, a positron emission tomography (PET) study found posterior medial OFC activation in normal subjects who performed a memory task requiring that they carefully separate relevant from similar but irrelevant memories (Schnider et al. 2000). Imaging studies by Schacter and colleagues support Schnider's claims, since they also show activation in the orbitofrontal cortex during recall, especially when a false memory is produced. Schacter et al. suggested that "blood flow increases in prefrontal cortex and cerebellum reflect subjects' efforts to inhibit … monitor … or make decisions about the sense of familiarity or recollection associated with false recognition" (1996a, 272). In a more recent article, the group speculated that "verification processes are particularly demanding in the case of False items, because they elicit a vivid recollective experience but less sensory recovery than True items" (Cabeza et al. 2001, 4809).

Domain Specificity
The question of the domain specificity of confabulation becomes important in assessing and improving different theories of confabulation. Several cases have been reported of patients whose confabulation was restricted to autobiographical memory, or who at least confabulated mainly about autobiographical memory (Nedjam et al. 2000). There are exceptions: Kopelman et al. (1997) described a patient who showed spontaneous confabulation across episodic, personal, and general semantic memory, whom they compared with another patient with frontal damage who only showed confabulation in "personal" semantic memory—memory of the important facts of one's life.

Whereas retrieval theories are supported by confabulation restricted to certain memory domains, executive or reality monitoring theories have trouble with domain-specific confabulation because the general understanding of executive processes is that they can be applied in more than one domain. Executive processes may have their own domain specificities, however. If there are several different checking or monitoring processes, this predicts that there are several different types of confabulating patients, depending on which processes are damaged. This also implies that several different neurological tests have to be used to assess each type.

Dalla Barba and colleagues (1990) developed a confabulation battery that contains questions touching on personal semantic memory (one's address, age, or occupation), autobiographical memory, orientation in time and place, and general semantic memory, such as historical facts. Inspired by Mercer et al. (1977), Dalla Barba's group included several extremely difficult questions in their battery (e.g., "Who was Marilyn Monroe's father?"), to which normal subjects would be likely to answer "I don't know," on the assumption that confabulators would fail to know that they did not know (see chapter 8).

3.6   Separating Amnesia and Confabulation

The sentence "Confabulation involves memory problems" might be understood as making either a definitional claim or a factual one. We should let the facts of the matter guide our definition, however. We must keep separate the following three hypotheses. First, memory problems are sufficient for confabulation; that is, if a person has a memory problem, he will confabulate. This is disproved by the existence of medial temporal lobe amnesics. Second, memory problems are necessary for confabulation; that is, if a person confabulates, he has a memory problem. This is the primary claim of those who hold a mnemonic conception of confabulation. Finally, there is the weaker claim that memory problems are correlated with confabulation. No one denies this, but the question is how high the correlation is.

A large body of evidence shows dissociations between problems with temporal memory and confabulation. Kopelman (1989) observed that amnesic patients and those with Alzheimer's disease frequently have problems with the temporal order of their memories without showing confabulation, including all those with Korsakoff's syndrome in his 1989 study. Patients with frontal lobe lesions can similarly show impairment of temporal context memory without confabulation (Shimamura et al. 1990). There are also ACoA aneurysm patients with memory impairment who do not confabulate (Vilkki 1985). In their study of confabulating patients with Alzheimer's disease, Nedjam et al. (2000) flatly state that memory deficits are neither necessary nor sufficient to produce confabulation. The two phenomena also show different time courses. Confabulation in Korsakoff's syndrome often clears, while the amnesia remains (Stuss et al. 1978). Similarly, confabulation clears in most patients with ACoA syndrome after weeks or months, but memory problems do not improve (Kapur and Coughlan 1980). Sometimes it works in the other direction; memory improves while confabulation continues (Stuss et al. 1978).

Based on results of a study of a patient who confabulated across autobiographical and semantic memory, Kopelman et al. (1997), agreed with Johnson that spontaneous confabulation cannot be explained entirely by either frontal dysfunction or impairment of temporal context memory alone. Talland said that the amnesia alone in Korsakoff's syndrome cannot explain the confabulation, for two reasons: "First, confabulation tends to disappear as patients progress from the confusional stage to the chronic phase, while the amnestic disturbance remains constant or changes only slightly. Second, some chronic patients confabulate more, others less or not at all, and there is no discernable correlation between their propensity to confabulate and the gravity of their amnesia" (1965, 44). DeLuca (1993) and Stuss et al. (1978) also argued that confabulation and memory problems are caused by damage to different neural systems.

Nevertheless, a correlation exists between memory problems and confabulation. In a study comparing nonamnesic and amnesic patients with ACoA syndrome, only the amnesic patients confabulated (DeLuca 1993). Cunningham et al. (1997) divided 110 patients with "diverse neurologic and psychiatric diagnoses" into high-, low-, and nonconfabulation groups based on the number of confabulations they produced in a story recall test. High-confabulation patients were significantly worse than the low- and nonconfabulation groups in other memory tests. Schnider et al. (2000) also showed that recovery of temporal confusion in memory paralleled recovery of confabulation.

If memory confabulation results from two independent lesions, this indicates that there are two types of patients:

1. Patients who sustain the memory system lesion first   They should admit their memory problem until the executive problem develops, at which point they should deny it, and begin confabulating.
2. Patients who sustain the executive lesion first   The course of this disease might be rather subtle. We also have to leave open the possibility that some people do not develop the executive processes necessary to check memory reports, and make do with their memories alone, tolerating a high rate of errors. This opens up the interesting possibility that the problem with some patients with Korsakoff's syndrome is that they are confabulatory before they lose their memory. They have lost the ability to check thoughts or candidate memories. It may pass unnoticed because the patient is substantially correct in what he or she says. But once amnesia sets in, the problem becomes painfully obvious.

3.7   False Memories

Apparently the memory systems themselves have a certain basic accuracy level, but then we use different frontal checking procedures to increase this level. As I noted in the introduction to this chapter, normal correct memories are rational reconstructions in that the reconstruction process is guided by what seems rational to the person. One can see this in certain patterns of error in false memories, where something odd in an event is misremembered as something more normal or rational. The phrase "false memory" is a bit of a contradiction in terms, of course, given that it is reasonable to hold that something is not a memory of any sort if it is not correct, but its meaning is clear enough.

Johnson and colleagues made the interesting observation that children exhibit some of the same patterns of memory problems that patients with frontal damage show (Lindsay et al. 1991). This may be because the frontal lobes are among the last cortical areas to mature. A large part of their development occurs between five and ten years (Case 1992), and they do not fully mature until the teenage years (Smith et al. 1992). An experiment similar to the one conducted by Ackil and Zaragoza (1998) and described in chapter 1, in which false memories were induced in children by asking them leading questions, was conducted by Ceci et al. (1994). Preschool children were presented with a deck of cards, each of which described an event. Some of the events had actually happened to the children and others had not. When they were repeatedly asked whether the false events had happened to them, fifty-eight percent of the children eventually agreed that they had, and many of them embellished the event with confabulated details.

There are also ways to elicit false memories in adults. Deese (1959) presented normal subjects with a list of words related to sleep, excluding the word "sleep" itself: "bed," "rest," "awake," "tired," "dream," "wake," "snooze," and so on. When they were later tested, between thirty and forty percent of the subjects claimed that they had seen the word "sleep." The brains of normal subjects were observed using PET as they performed a Deese-type task in which the subjects first heard a list of related words and then were tested for memory of the words. Both true positives and false positives involved activation of hippocampal areas, but true positives showed additional activation of auditory cortical areas (Schacter et al. 1996a). The authors speculated that this additional activation may involve perceptual representations of spoken words.

It may be that nature's plan is that the checking processes will be instilled after birth, during the long training period we humans require, principally by our parents. What begins as an external loop is made internal; the child confabulates, the parent corrects, the child changes what he said. We then internalize these corrections so that the loop runs completely within our brains, although it shares some of the same dynamics. There is still a candidate claim, and there is still a check that has the power to inhibit the claim from being made.

3.8   Conclusion

The need for a causal connection of the appropriate sort between a memory and the event it represents seems to be met in some cases of confabulation, but the patient is wrong about when the event happened. One question about a memory that is displaced in time, however, is whether the memory was recalled with the wrong time represented, or whether this was added by the patient in the process of reporting the memory. In either case, the existence of confabulation makes it clear that memories with irretrievable time tags are of very limited use to us.

Delbecq-Derousne et al. (1990) studied a patient with ACoA syndrome who showed what they called pathological false recognition. In a forced-choice experiment in which he had to indicate whether he had seen a stimulus before, the patient produced a large number of false positives, and his confidence ratings for those answers were not different from those given for his correct responses. The confidence of confabulating patients often makes one wonder whether it is based not so much on a negative phenomenon (the absence of a feeling of doubt) as on a positive phenomenon. William James (1890) suggested that we distinguish memories from mere images by the "feeling of warmth" that accompanies memories. If there is such a thing, it is plausible that its presence might fool someone into believing a false memory was real.

Subjects of hypnosis will also confabulate about memories. Dywan (1995) identified two independent hypotheses for why hypnotized people confabulate: They lower the criteria they use in evaluating memory reports, or they experience an illusion of familiarity. Dywan preferred the second approach because "hypnosis alters the experience of retrieval" so that what is retrieved is more likely to have qualities such as vividness and "perceptual fluency" (1995, 194). In fact, this hypothesis explains the high levels of confidence that hypnotized subjects have in their confabulated answers; thus confabulation is at least facilitated by a "positive" phenomenon, an increase in the vividness of apparent memories.

Neurologist Alexander Luria, with his uncanny ability to point to the source of a problem, anticipated many of the themes of this chapter in a 1978 article. He mentioned the trouble keeping irrelevant memories at bay that Schnider focused on, as well as the patients' disorientation, which he cited as the cause of confabulation. When the frontal lobes are damaged, said Luria, the cortex is

unable to distinguish between strong, dominant foci of excitement, and foci induced by weak, interfering stimuli. The distinction between important and irrelevant foci disappears [and] a disturbance of the selectivity of connections arises. The patient loses the precise orientation in space and, in particular, in time; he considers that he is in some indefinite place: "in the hospital," "at work," or "at the station." Sometimes the primary disturbance of orientation is compensated by naïve uncontrollable guesses: seeing the white gowns and white hats of physicians, the patient declares that he is "at the baker's'," "at the barber's." Such a patient cannot answer when he is asked where he was the same morning or night before, and the irrepressible traces of past experience lead him to fill this gap with confabulations: he says he was "at work," or "walking in the garden," and so on. The contradiction between his assessment and the real situation causes such a patient little confusion because the rapidly disappearing traces of his impressions do not provide an opportunity for reliable, critical, comparisons. (Luria 1978, 26–27)

The last remark, about the way that his impressions rapidly disappear, is on target in that confabulators seem to have a problem holding clear representations in consciousness so that they can be assessed by frontal processes (see chapter 9). This idea also helps to explain the way they impassively admit the truth when confronted with it, then minutes later lapse back into confabulations.

If Korsakoff's syndrome involves damage to the thalamic mediodorsal nucleus, and aneurysm of the anterior communicating artery often involves damage to the orbitofrontal cortex, understanding the circuit that connects these two areas is important for understanding confabulation. This seems to be our best clue thus far toward solving the riddle of confabulation. The orbitofrontal cortex and its connections are thus the foci of chapter 4.

4 Liars, Sociopaths, and Confabulators

The reflective man makes use of his critical faculties, with the result that he rejects some of the thoughts which rise into consciousness after he has become aware of them, and abruptly interrupts others, so that he does not follow the lines of thought which they would otherwise open up for him; while in respect of yet other thoughts he is able to behave in such a manner that they do not become conscious at all—that is to say, they are suppressed before they are perceived.
—Sigmund Freud (1900/1938)

4.1 Unhappy Family: The Orbitofrontal Syndromes

Understanding confabulation is in one respect similar to the attempt to properly classify a recently discovered animal species—it must be placed in its proper family. In this chapter I introduce some other members of a family of dysfunctions to which I propose confabulation belongs, including sociopathy and a condition called disinhibition. The clues thus far are pointing us in the direction of the orbitofrontal cortex. Once we focus on it as a damaged area in confabulation, a look at other syndromes thought to involve orbitofrontal lesions confirms that we are in the right place. What we have to understand is the nature of the functions the orbitofrontal cortex is performing; not so much how we think of them, but how the brain thinks of them, as it were.

Chapter 3 cited several reasons for delving further into the structure and function of the orbitofrontal cortex. Both aneurysms of the anterior communicating artery and Korsakoff's syndrome may involve damage to the orbitofrontal cortex directly, or to areas that are densely interconnected with it, such as the thalamic mediodorsal nucleus in the case of Korsakoff's syndrome. Memory confabulation theorists also tend to point to the orbitofrontal cortex as a problem area. Johnson et al. say that "damage to the orbitofrontal circuit may disrupt processes (either inhibitory or excitatory agenda maintenance processes) that help foreground stimuli related to current agendas and the information that is most relevant to task goals" (2000, 378). Schnider's group also points at the orbitofrontal cortex as a problem area in confabulation, and Benson et al. (1996) argue that the orbitofrontal cortex is important for self-monitoring and self-awareness. Johnson et al. suggest likewise that anterior cingulate and orbitofrontal circuits "may be critical to a person's ability to self-monitor—that is, to take mental experiences (or behavior) as objects of reflection" (2000, 378).

In addition, an intriguing link exists between sociopathy and confabulation through the orbitofrontal cortex. The sociopath tells lies with an ease and confidence that resembles the act of confabulation, even if the confabulator is not actually lying. Several researchers have pointed to the orbitofrontal cortex as a problem area in sociopathy, and patients have been discovered who became sociopathic after damage to that structure, a condition known as acquired sociopathy. Patients with damage to the orbitofrontal cortex can also become disinhibited, meaning roughly that brain processes give rise to mental events and behavior that would normally have been blocked or suppressed by other brain processes. Recall John Hughlings Jackson's idea that damage to higher-level areas releases responses generated by lower-level areas. Damage to the orbitofrontal cortex "may disrupt processes that withhold responses (e.g., based on familiarity) until better information revives" (Johnson et al. 2000, 378). We might see confabulations as resulting from disinhibition also; the same sort of damage that loosens the behavior of the disinhibited person loosens the tongue of the confabulator. Making a claim is itself a kind of action, which confabulators engage in too easily. In this conception, damage to the orbitofrontal lobes disinhibits candidate responses that had been generated in the past, but inhibited. This helps to explain the individual differences seen in disinhibited behavior. People differ in the types of behaviors they had been inhibiting before brain damage, and in the frequency with which they engaged in inhibition.

Memory researcher Arthur Shimamura (1995) suggested that the function of the frontal lobes in general is inhibition, or what he called "filtering." The frontal lobes causally contact representations created by other brain areas, and support different processes to select out defective representations. Damage to them produces disinhibition of different varieties—of action, speech, and emotion—depending on where it occurs. "One possibility," Shimamura says, "is that the prefrontal cortex controls information processing by enabling inhibitory control—specifically inhibitory control of activity in posterior cortical regions" (1995, 810). Shimamura made an important connection to social behavior, noting that his filtering theory "may also be useful in explaining the kinds of social disinhibition observed in patients with orbital prefrontal lesions.... Such patients exhibit a failure to inhibit inappropriate social behaviors (e.g., telling crude jokes, emotional rages)" (1995, 811). There is a specificity to this filtering: "The result [of damage to different areas of the prefrontal cortex] is different because specific frontal regions are connected to different areas in posterior cortex, and thus are filtering different aspects of cognitive function" (1995, 810). This fits well with the observation in chapter 3 that confabulation can occur selectively in different domains of memory.

A new movement that originated in cognitive neuropsychology in the 1980s is based on the idea that at certain strategic points, cognition and emotion must interact. Emotion plays a guiding function in these interactions, without which cognition is not only ineffective, but rendered useless (Damasio 1994). The effects of emotion seem to come primarily at the input and output points to the cognitive system; that is, the sets of representations we possess, as well as the operations for manipulating them. At the input point there may be occasions in which certain emotional reactions are required in order for us to conceptualize a thing or person correctly (see chapter 5 for more on this). At the output point, certain emotional processes function to keep us from acting on thoughts or intentions that are either ill-considered, or based on ill-grounded beliefs—we might call these inhibitory emotions. An emerging consensus proposes that the filtering that the orbitofrontal lobes perform involves emotions.

After perception-action cycles progress toward the front of the brain, they must loop back to activate more centrally located motor areas that ultimately send impulses down the spinal cord, causing our actions. At the point at which thoughts and intentions might become actions (including speaking), representational areas and executive processes exist that can evaluate thoughts and if necessary prevent them from causing actions. Before you actually do something, it would be prudent to review your action in various ways and perhaps, if you have the representational resources, run a simulation of it and attempt to discern whether it will progress as you imagine. You can imagine the initial situation, then create a *forward model* by letting the simulation run its course, based on your ability to represent events of this sort. At each stage you also need a way to discern the value the represented event has for you; that is, you need to be able to discern whether the simulation is progressing in the direction you desire. Thus memories of reward values from previous contexts in which these representations or actions were being considered are necessary for this modeling to occur. For actions, context is crucial. The same action that is a stroke of genius in one context can be suicidal in another.

Sometimes during this modeling process we experience an inhibitory impulse. Much of the time the inhibition is traceable to social factors; that is, the action would offend, annoy, anger, disturb, or hurt another person. Or it would simply cause the hearer to misunderstand our message. To know whether or not someone would be angered or pleased by something, I have to know about her desires, beliefs, and other attitudes; I must know where she is, what she is looking at, and how she represents those things. I also need to know how she will understand and react to the particular words I plan to use. Here our attitudes toward that person are also important; e.g., we are willing to anger or hurt our enemies. Or I may simply not care a whit about anyone, so that people's attitudes only rarely play a role in planning my behavior. I still have to run forward simulations of my contemplated actions, but for strategic rather than emotional-social reasons.

Lying is a complicated art, especially when the stakes are high. The order of the three abnormal conditions in the chapter's title is deliberate: first liars, then sociopaths, then confabulators. It represents a gradient in the amount of awareness the speaker has that what he is saying is false and in the amount of tension this creates in him. This awareness involves the possession of representations contrary to what the person claims, but also it often involves an accompanying emotion. Many writers have remarked on the effortlessness with which confabulators make their claims, describing them as fatuous and indifferent to their situation. Real lying, on the other hand, can sometimes be detected in normal people by the emotional responses it sets off. As pioneers of the polygraph learned long ago, situations can be set up in which a liar's knowledge can be detected indirectly. Intriguing links also are seen between lying and the ideas that frontal areas work to inhibit. I suggest that the emotional response produced by lying normally functions to actively stop a thought from causing intentions or actions. We may mark a thought as false using the same emotional tagging process we use to mark thoughts as dangerous or socially inappropriate, i.e. as having led to negative consequences in the past. In sociopaths, these emotional reactions are either blunted or absent, and in clinical confabulators the brain systems that produce these emotional reactions are damaged so that they are completely gone from the patients' mental life.

4.2   Symptoms of Orbitofrontal Damage

German psychiatrist Leonore Welt suggested in 1888 that damage to the orbitofrontal lobes could produce disinhibition. She described a patient who had damaged his orbitofrontal lobes when he fell from the fourth floor of a house. He was overtalkative, restless, and euphoric, and Welt noted that he showed a lowering of ethical and moral standards. Finger described the patient in his superb *Origins of Neuroscience*: "[B]efore his fall he had been a relaxed man of good humor. Afterward he became critical, aggressive and annoying. He blamed other patients and the hospital staff for his problems, played mean tricks on them" (Finger 1994, 273). After his death, autopsy revealed "a destruction of the right inferior frontal gyrus as well as damage to the gyrus rectus on both sides" (Finger 1994, 274; figure 4.1, number 12). Welt concluded from her research that these sorts of personality changes were most reliably caused by damage to the right medial orbital prefrontal lobe.

Phineas Gage
In the summer of 1871, a railroad crew foreman was using a long steel rod to softly compress a gunpowder charge in a hole drilled by his work crew. As an experienced blaster, Phineas Gage knew that this "tamping iron" had

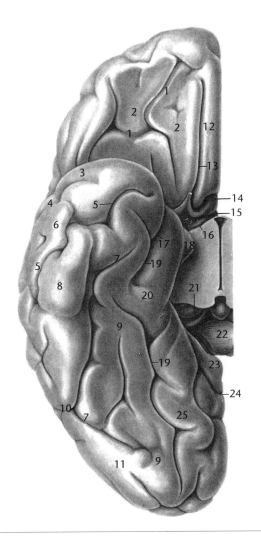

1	Sulci orbitales	14	Area subcallosa
2	Gyri orbitales	15	Gyrus paraterminalis
3	Gyrus temporalis superior	16	Gyrus diagonalis
4	Sulcus temporalis superior	17	Sulcus rhinalis
5	Sulcus temporalis inferior	18	Gyrus ambiens
6	Gyrus temporalis medius	19	Sulcus collateralis
7	Sulcus occipitotemporalis	20	Gyrus parahippocampalis
8	Gyrus temporalis inferior	21	Pulvinar thalami
9	Gyrus occipitotemporalis lateralis	22	Splenium corporis callosi
10	Incisura preoccipitalis	23	Isthmus gyri cinguli
11	Gyri occipitales	24	Sulcus calcarinus
12	Gyrus rectus	25	Gyrus occipitotemporalis medialis
13	Sulcus olfactorius		

Figure 4.1
The ventral surface of the right hemisphere. The main gyri and sulci of the orbitofrontal lobes are visible at the top. (From Nieuwenhuys et al., 1988, with permission.)

to be used gently before filling the hole with sand, but his attention was distracted by his crewmen for a moment. As his physician, John Harlow told it: "Averting his head and looking over his right shoulder, at the same instant dropping the iron upon the charge, it struck fire upon the rock, and the explosion followed, which projected the iron obliquely upwards ... passing completely through his head, and high into the air, falling to the ground several rods behind him, where it was afterwards picked up by his men, smeared with blood and brain" (1869, 275). His men put Gage in an oxcart and took him back to his hotel, where Harlow was contacted. Gage suffered loss of vision in his left eye as well as some paralysis of the left side of his face, but survived the blow and subsequent infection, and ultimately regained his health. What had changed greatly, however, was his personality—so much that his former employers had to fire him shortly after he returned to work.

Harlow's diagnosis of Gage was that the "equilibrium or balance, so to speak, between his intellectual faculty and animal propensities, seems to have been destroyed. He is fitful, irreverent, indulging at times in the grossest profanity (which was not previously his custom), manifesting but little deference for his fellows, impatient of restraint or advice when it conflicts with his desires, at times pertinaciously obstinate, yet capricious and vacillating, devising many plans of future operation, which are no sooner arranged than they are abandoned in turn for others appearing more feasible" (Harlow 1869, 277). His "manifesting but little deference for his fellows" qualifies as impaired social behavior one would think. These were big changes: "Previous to his injury ... he possessed a well-balanced mind, and was looked upon by those who knew him as a shrewd, smart businessman, very energetic and persistent in executing all his plans of operation. In this regard his mind was radically changed, so decidedly that his friends and acquaintances said he was 'no longer Gage'" (Harlow 1869, 277). Gage seems to have become disinhibited in many realms, including making claims: Harlow said that Gage "was accustomed to entertain his little nephews and nieces with the most fabulous recitals of wonderful feats and hair breadth escapes, without any foundation except in his fancy" (1869, 277). Damasio's (1994) reconstruction of the injury revealed that it destroyed posterior medial orbitofrontal cortex, as well as large portions of medial prefrontal cortex and the anterior cingulate cortex located above it, disturbing the functions of the areas shown in figure 4.2.

In 1975 Blumer and Benson described a similar patient who had most of his right orbitofrontal cortex removed after an infection caused by a bullet wound he sustained while serving in Vietnam. "Prior to his injury he had been quiet, intelligent, proper, and compulsive. He was a West Point graduate and spent the ensuing years as a military officer attaining the rank of captain. Both as a cadet and later as an officer, he was known to be quiet,

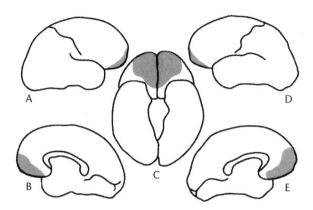

Figure 4.2
Damasio's diagram, showing the parts of the frontal lobes that are damaged in the Gage matrix. (From Damasio, 1994, with permission of the author.)

strict, and rigid. Subsequent to his injury, he was outspoken, facetious, brash, and disrespectful" (1975, 155–156).

Experimental brain surgery, first on animals, then on humans, affirms the idea that frontal lobe damage can cause a disinhibition that strongly affects the subject's social interactions. Leonardo Bianchi (1922) focused on changes in what he called the "sociality" of the dogs and monkeys he studied. He noted that damage to the prefrontal lobes affected emotions that guide our interactions with others. "Bianchi singled out the social sentiments (friendship, social understanding, true love, social obligation, etc.) as those emotions most clearly dependent on the frontal lobes" (Finger 1994, 275).

Prefrontal Lobotomy
In 1935, Egas Moniz, a Portuguese professor of neurology, performed the first prefrontal lobotomy after hearing of a monkey who went from being agitated and volatile to looking as if it "had joined the 'happiness cult of the Elder Micheaux,' and had placed its burdens on the Lord!" after frontal lobe ablation (Finger 1994, 291). In the United States, Walter Freeman and James Watts began performing the operation a year later, primarily in an attempt to help patients with severe obsessive-compulsive disorder. Of interest to us is that they explicitly attempted to sever the connections between the mediodorsal thalamus and the prefrontal cortex. They said that "the dorsal medial nucleus of the thalamus by its connections with the cerebral cortex may well supply the affective tone to a great number of intellectual experiences and through its connections with the hypothalamus

may well be the means by which emotion and imagination are linked" (Freeman and Watts 1942, 27).

Freeman and Watts noted the following characteristics in patients after a frontal lobotomy: inertia, lack of ambition, a decrease in "consecutive thinking," indifference to the opinions of others, and satisfaction with results of inferior quality. Lobotomy also produced what Hecaen and Albert summarized as "reduction of affective response inhibition, tendency to euphoria, emotional instability, and impairment of attention" (1975, 141). Stengel (1952) studied hundreds of psychosurgery patients and noted that many of them denied having had the operation and seemed detached and unconcerned about their situation (see chapter 6). Lobotomy also caused what Stuss called "a disturbance in the capacity of foresight in relation to the self" (1991, 66). "The lobotomized patient," said Stuss, "has lost awareness of himself as a continuing and changing entity, with personal responsibility for such a change" (1991, 66). Rolls similarly mentioned that ventral frontal patients who had been in accidents and been awarded compensation "often tended to spend their money without appropriate concern for the future" (1999, 131).

Hecaen and Albert also spoke of a "simplification of social interactions" in lobotomized patients (1975, 142). In a review of the effects of lobotomy, Greenblatt et al. (1950) described a decreased social sense. "[Orbitofrontal cortex]-lesioned patients are often described as coarse, tactless, and generally lacking in empathy and social restraints," say Zald and Kim (2001, 53), who add the following interesting remark: "Despite an apparent disregard for social rules, they usually lack the intentional viciousness or organization of a true antisocial personality disorder" (2001, 53). In general, lobotomy and the other types of lesion to the orbitofrontal cortex have a disinhibiting effect. What appears in the patient's behavior is a product both of the disinhibition and processes that gave rise to the impulses that are acted upon. Since people differ in the types and numbers of impulses that they have, the results of this disinhibition vary from person to person.

Damasio's E.V.R.

In his fascinating book, *Descartes' Error* (1994), neuropsychologist Antonio Damasio described a patient, E.V.R., who appeared to have fully recovered from an operation to remove a brain tumor that had damaged the ventromedial portions of his orbitofrontal cortex. Soon after the operation, however, bad things began to happen to E.V.R.: several businesses failed; he divorced his wife, remarried, and then quickly divorced again. His problem seemed to be that he was making poor decisions. Damasio gave E.V.R. all of the relevant tests of brain dysfunction, but he scored well on them. Finally, it occurred to Damasio that the problem might be emotional rather than cognitive. He and his colleagues (see Bechara et al. 1994) monitored auto-nomic activity (in the form of skin-conductance responses, or SCRs) of patients with brain damage similar to that of E.V.R. while they played a gambling game in which the player must choose between selecting a card from a "safe" deck, which allows small but steady gains in winnings, or from a "risky" deck, which offers large winnings but even larger losses. After several exploratory trials, normal subjects showed an SCR when they reached for the risky deck, and they soon learned to stay with the safe one. Brain-damaged patients, however, did not register this SCR and did not learn to avoid the risky deck. Damasio's conclusion: E.V.R.'s problem was that although he often knew intellectually what the right course of action was, the lack of an emotional reaction to certain thoughts and situations caused him to continue engaging in dangerous or harmful activities.

Damasio and his colleagues described the process by which normal people use a combination of emotion and memory to gauge the desirability of a contemplated action: "When subjects face a situation for which some factual aspects have been previously categorized, the pertinent dispositions are activated in higher-order cortices. This leads to the recall of pertinently associated facts which are experienced in imagetic form. At the same time, the related ventromedial prefrontal linkages are also activated, and the emotional dispositions apparatus is competently activated as well. The re-sult of these combined actions is the reconstruction of a previously learned factual-emotional set" (Bechara et al. 2000a, 297).

Activation of these emotions can be accompanied by activation of the autonomic components of emotion, what Damasio called the body loop. Or the activation may occur by the shorter "as-if body loop," where the representations of the viscera (e.g., in the parietal cortex) are altered to fit the emotion currently being felt. Of interest for our purposes is the fact that many patients with ventromedial damage who show a deficit in the gambling task have suffered ACoA aneurysms (Bechara et al. 1996). Mavaddat (2000) found specifically that a group of patients with ACoA syndrome did indeed fail the Bechara–Damasio gambling task. The task also involves sensitivity to probabilities, which may be related to the errors in cognitive estimation reported in patients with ACoA syndrome (Beeck-mans et al. 1998).

Rolls's Theory

Edmund Rolls's book *The Brain and Emotion* contains a comprehensive look at brain anatomy and function related to emotion. Rolls based his approach primarily on experimental work with rhesus monkeys and other animals. He described orbitofrontal damage in general as producing "a failure to react normally to and learn from non-reward in a number of dif-ferent situations," which produces "a tendency to respond when responses are inappropriate" (1999, 115). The orbitofrontal cortex is "involved in correcting behavioral responses made to stimuli previously associated with reward" (1999, 116; see also Jones and Mishkin 1972). The most reliable finding about the behavior of monkeys with orbitofrontal lesions is that they tend to keep responding long after a response has ceased to be rewarding.

The orbitofrontal cortex seems to be able to encode the reward value of a stimulus even if that value changes rapidly. For instance, orbitofrontal neurons connected to visual areas reverse their responses to visual stimuli when the taste paired with the object is changed (Rolls 1999). Patients with ventral frontal damage also made more errors in a task in which reward contingencies were changed (Rolls et al. 1994). Rolls says that "the impairment correlated highly with the socially inappropriate or disinhibited behavior of the patients" (1999, 131). This ability to reverse the representation of reward value associated with a stimulus "is important whenever behavior must be corrected when expected reinforcers are not obtained, in, for example, feeding, emotional, and social situations" (Rolls, 1999, 120). Rolls criticized Damasio's approach as being too roundabout: "Somatic markers are not part of the route by which emotions are felt or emotional decisions are taken, but ... instead the much more direct neural route from the orbitofrontal cortex and amygdala to the basal ganglia provides a pathway which is much more efficient" (1999, 138).

Disinhibition

Another piece of evidence that confabulation involves problems in the front of the brain is that it is often accompanied by disinhibition, which is normally the product of frontal damage. Disinhibition comes in a variety of forms. The basic premise is that the patient engages in behaviors that might have occurred to him but would have been rejected before the brain damage. Starkstein and Robinson (1997) distinguished five types of disinhibition: (1) motor disinhibition can cause hyperactivity, "pressured speech," and a decreased need for sleep; (2) disinhibition of instincts can cause hypersexuality, overeating, and aggressive outbursts; (3) emotional disinhibition can cause the patient to express fully whatever emotion he is feeling at the moment, sometimes going from crying to laughing within seconds; (4) intellectual disinhibition can cause the patient to voice grandiose or paranoid delusions; and (5) sensory disinhibition can produce visual and auditory hallucinations. They note that "although patients with brain lesions may sometimes show a disinhibition syndrome restricted to one of the above categories, most frequently they present with a mixture of them" (1997, 108).

A disinhibited person may make socially inappropriate remarks or constantly use profanity, as Gage did. Fukutake et al. describe a disinhibited patient who had an infarct in the "ventroposteromedial part" of the left thalamic mediodorsal nucleus, perhaps indicating that disruption of some of the orbitofrontal cortex's connections can also cause disinhibition. Their patient, a forty-eight-year-old man was emotionally unstable after the infarct. He showed "explosiveness or lability (he made many demands to go to the toilet, take a meal, leave the hospital and would not wait for permission from the medical staff nor control outbursts of rage and sometimes of crying" (2002, 158). He was also "distractible and had poor concentration (subjects of his conversations often changed)" as well as a "poverty of flexible thinking and tact (he would stubbornly stick to his demands)" (2002, 158).

Starkstein and Robinson (1997) offered an interesting theory of the anatomy of disinhibition based on the relationships between the dorsal and ventral cortical moieties (see section 2.2). They suggested that the dorsal trend functions to produce behaviors, whereas the ventral cortical trend, and especially the orbitofrontal cortex, function primarily to either inhibit or release candidate behaviors produced by the dorsal trend. They also suggested that when orbitofrontal damage releases dorsal areas from ventral modulation, both intellectual and sensory disinhibition may result. In their theory, damage to either the orbitofrontal cortex or its connections to the striatum, which are thought to be important in the genesis of voluntary action, may produce motor disinhibition. They suggested that instinctive disinhibition may occur when the orbitofrontal cortex is disconnected from more primitive brain structures involved in drives, such as the hypothalamus and amygdala.

In the view I set out in chapter 8, the disinhibition of different types present in the orbitofrontal syndromes is an important clue to understanding confabulation. Confabulation should be seen as a type of disinhibited action. Sometimes we forget to see making claims as an action, perhaps because we associate it more with thought than with action. To speak to someone is to direct an action at the person just as much as to reach out to shake hands. Just as with shaking hands, the other person has to be aware that the gesture has been made, and has to reciprocate. We might view confabulation as a type of epistemic disinhibition—the patient makes knowledge claims he or she would not have made before the brain damage.

Damage to the orbitofrontal lobes on occasions seems to damage the patient's standards of quality. The claims of the confabulator are of an inferior quality, we might say. Recall that Freeman and Watts (1942) noted that lobotomized patients were satisfied with results of inferior quality. Logue et al. described a fifty-eight-year-old man who had an ACoA aneurysm. According to his wife, "he used to be moody and sulky and fault-finding—now he is more mellow in his ways" (1968, 141). The epistemic standards of quality are also damaged in the confabulator.

4.3   Anatomy and Physiology of the Orbitofrontal Cortex

The orbitofrontal cortex is a large area, but we have information that can help narrow our focus. We have special reason to be interested in those parts of the orbitofrontal cortex that are either interconnected with the thalamic mediodorsal nucleus, a known site of damage in Korsakoff's syndrome, or are likely to be damaged by an aneurysm of the ACoA, i.e., the more posterior and medial portions. We have examined some possible functions of the orbitofrontal cortex, but one great value of neuroanatomy is that it can tell us which other brain areas the orbitofrontal cortex interacts with, and together the neurophysiologist and the cognitive neuropsychologist can tell us what the functions of those areas are thought to be. We can thus develop a sort of functional map (see section 2.2) of the brain systems that are damaged in confabulators.

The orbitofrontal cortex is the highest-level area of the ventral cortical trend, and one of the ultimate destinations for the ventral visual processing stream—the stream that functions primarily to allow us to detect, categorize, and associate reward values with objects and people (see section 2.2).

Figure 4.3 depicts the classical Brodmann areas composing the orbitofrontal cortex and the adjacent medial prefrontal cortex. The orbitofrontal cortex is adjacent to and in front of the motor and premotor cortices. It receives visual input from the superior temporal sulcus, inferior temporal cortex, and temporal pole. It also receives auditory input (Barbas 1993) and somatosensory input (Carmichael and Price 1995b).

Neuroanatomists Carmichael and Price (1996) discovered two large, relatively segregated processing networks in the orbital and orbitomedial prefrontal cortices of the rhesus monkey (figure 4.4). The *medial network* interconnects certain orbital areas with medial areas. It involves primarily limbic structures and is well situated to produce autonomic responses. "The medial orbital cortex ... stands out as a site of convergent amygdaloid, hippocampal, and other limbic inputs to the prefrontal cortex" (Carmichael and Price 1995a, 636). In contrast, the *orbital network* links most parts of the orbitofrontal cortex but has limited connections to medial areas. The orbital network receives sensory input from the dorsolateral prefrontal cortex and from visual, somatosensory, gustatory, visceral, and olfactory areas. The two networks share certain nodes located in areas 13a and 12o (figure 4.4), which Carmichael and Price suggested may mediate interactions between them. Since it is in the posterior portion of the orbitofrontal lobes at the posterior end of the gyrus rectus, area 13a is vulnerable to damage caused by an aneurysm of the anterior communicating artery.

As a whole, the orbitomedial prefrontal cortex "acts to link environmental stimuli and experience with somatic or visceral responses, and at

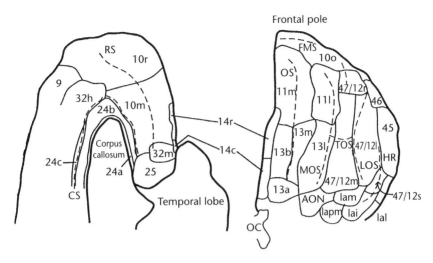

Figure 4.3
Map of the (left) human prefrontal cortex. Area 47 in the human brain corresponds to area 12 in the rhesus monkey brain. Medial surface on the left, orbital surface on the right. Sulci are indicated by dashed lines. (From Ongur, D. and Price, J. L. 2000. The organization of networks within the orbital and medial prefrontal cortex of rats, monkeys, and humans. *Cerebral Cortex* 10: 206–219, with permission.)

the same time pairs them with the appropriate mood or affect" (Carmichael and Price 1996, 206). In their study of the connections between the orbitofrontal cortex and the autonomic centers in the hypothalamus, Ongur et al. describe the functional relations between the two networks: "Sensory information from several modalities related to motivational or emotional aspects of ongoing behavior ... is sent to the orbital network where it is processed in hierarchical fashion. The salient features of this information are then transferred to the medial network, which elicits an appropriate visceral reaction by modulating activity in the hypothalamus ... and related structures such as the amygdala" (1998, 504).

All of these areas work together to relate emotion to memories of events: "Based on the role of the amygdala, and the [orbitofrontal cortex] in autonomic and emotional function, it can be hypothesized that the anterior limbic-prefrontal network is involved in affective responses to events and in the mnemonic processing and storage of these responses, including stimulus-reward association and reversal" (Carmichael and Price 1995a, 639). Pandya and Yeterian provide a summary of the roles of the orbital and orbitomedial prefrontal regions in emotions: "These regions play a role

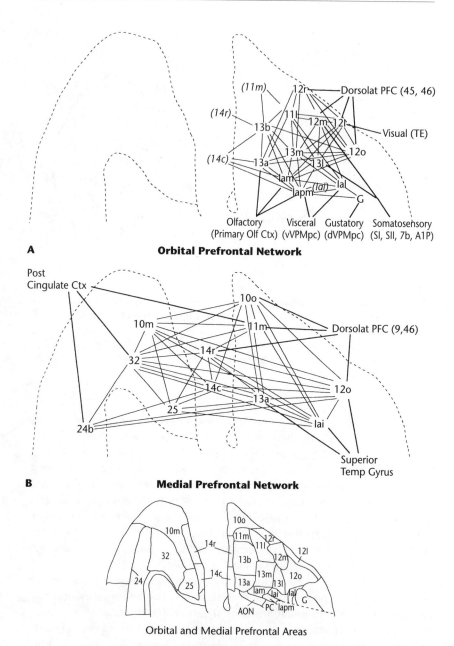

A   **Orbital Prefrontal Network**

B   **Medial Prefrontal Network**

Orbital and Medial Prefrontal Areas

Figure 4.4
Diagram of the orbital and medial networks in the rhesus monkey brain. Notice
the inputs from the superior temporal gyrus to the medial network. (From
Carmichael, S. T. and Price, J. L. 1995b. Sensory and premotor connections of
the orbital and medial prefrontal cortex of macaque monkeys. *Journal of Comparative Neurology* 363: 642–664, with permission.)

in internally generated mood, in the appreciation of emotional facial and vocal expressions, in decision making in relation to autonomic function, in emotional planning, in emotional empathy, in the appreciation of pleasant somatosensory and auditory (musical) stimuli, and in the appreciation of aversive olfactory and gustatory stimuli" (2001, 78).

Thalamic Connections
Projections from the magnocellular division of the mediodorsal nucleus (figure 3.1, number 4 and figure 3.2, number 5) to orbital areas 11, 12, and 13 are reciprocated by significant projections back to the same parts of the thalamus. Parts of the mediodorsal nucleus that project to the medial and orbital networks are segregated within the mediodorsal nucleus. The medially projecting parts "are clustered in the dorsomedial and caudal parts of the nucleus" (Carmichael and Price 1996, 204), whereas portions that project to the orbital network are in the ventromedial part of the mediodorsal nucleus. The more medial parts of the orbitofrontal cortex also receive projections from the anteromedial nucleus of the thalamus (Carmichael and Price 1995a). Recall from chapter 3 that damage to the human anterior thalamic nuclei was detected in an imaging study of patients with Korsakoff's syndrome. The intralaminar thalamic nuclei, thought to play a role in producing conscious states (e.g., Llinás 2001), and the midline thalamic nuclei also project to the medial orbital cortex (Barbas et al. 1991).

In their study of orbitofrontal connections to the mediodorsal nucleus, Ray and Price noted a function the mediodorsal nucleus (MD) has in inhibiting activity: "If thalamocortical activity provides a mechanism for sustaining activity related to concentration on a particular task, inhibition of MD may block sustained activity, in order to switch between different patterns of thalamocortical activity, or to suppress unwanted thoughts or emotions" (1993, 29). In the better-understood motor system it is well confirmed that the globus pallidus functions to suppress inappropriate activity in antagonistic muscle groups during movements. The mediodorsal nucleus has "a similar role, but in relation to cognitive or affective functions instead of motor behavior" (Ray and Price 1993, 29). If inhibition of the mediodorsal nucleus can suppress certain thoughts or emotions, perhaps damage to it has a similar effect, which might explain the general tendency of confabulating patients to be bland, serenely calm, and untroubled by information contrary to what they claim.

Amygdala Connections
The orbital and medial areas are the only prefrontal areas with strong amygdala connections (Zald and Kim 2001). LeDoux (1995) referred to the amygdala as an "emotional computer." Its general role may be to signal danger (LeDoux 1995); it has been found to respond to the sight of another

person showing fear, for instance. The amygdala is composed of several distinct nuclei that are extensively interconnected. Among the areas we are interested in, the amygdala is primarily connected with the medial network of Carmichael and Price, although its connections include area 13a, which is part of both the medial and orbital networks. The amygdala and orbitofrontal cortex are similar in that they both receive input from high-level sensory areas, and they are both capable of producing autonomic reactions through their connections to the hypothalamus and other autonomic centers.

Epileptologists who stimulate the brains of patients with epilepsy before surgery found that one of the surest ways to produce a conscious experience is to stimulate the amygdala. This need not imply, however, that the amygdala is the place where conscious emotional experience takes place, but it may mean that it has close connections to such areas. One piece of evidence against the idea that emotional experience takes place in the amygdala itself is the fact that fearful faces that are masked in such a way that subjects do not consciously perceive them nevertheless produce amygdala activation (Whalen et al. 1998). Papez declared in 1937 that, "for subjective emotional experience ... the participation of the cortex is essential" (Finger 1994, 287), and this is still widely believed. The necessity of cortical participation raises the interesting possibility that some part of the conscious experience produced by stimulation of the amygdala is due to activation of the orbitofrontal cortex.

Hypothalamic Connections

The hypothalamus (see figure 3.4) is a major coordinating center of the autonomic nervous system. As we will see in detail in the next section, damage to the orbitofrontal cortex in humans can bring about failure to produce an autonomic response to certain types of emotional stimuli, particularly those that produce negative emotions. The hypothalamus regulates blood pressure changes, pulse, respiration, perspiration, water intake and output, and many aspects of sexual activity. It integrates information from several different sources, including the orbital and medial cortical areas we are interested in, as well as the thalamus. Stimulation of the orbitofrontal cortex can produce autonomic responses, presumably through connections to the hypothalamus. In 1949, Kaada et al. discovered that electrical stimulation of the posterior orbitofrontal cortex caused changes in blood pressure, heart rate, and respiration rate in monkeys.

Rempel-Clower and Barbas (1998) performed a staining study to delineate which prefrontal (including orbitofrontal) areas had connections to the hypothalamus. They found that the incoming and outgoing connections converged in the hypothalamic areas known to have autonomic functions. The orbitofrontal cortex is reciprocally connected to the lateral hypothalamus, which is a major output center for the limbic system as a whole. The lateral hypothalamus sends diffuse projections to many parts of the orbitofrontal cortex, but the connections running the other way originate primarily from the posterior and medial parts of the orbitofrontal cortex; that is, from the parts that participate in the medial prefrontal network (Ongur et al. 1998).

Electrical stimulation of the orbital cortex can produce both sympathetic and parasympathetic activation, but damage to the orbital cortex does not affect the tonic aspects of autonomic functions (Zald and Kim 2001), that is, baseline activity as opposed to momentary or phasic bursts of sympathetic activity. Recall from chapter 3 that the anterior portions of the hypothalamus are fed by the anterior communicating artery. This would include the preoptic region of the hypothalamus, which is also interconnected with the orbitofrontal cortex (Rolls 1999).

Anterior Cingulate Connections

The medial orbitofrontal cortex is also densely interconnected with both the anterior and posterior cingulate cortices (Vogt and Pandya 1987). As we saw in chapter 3, the anterior cingulate is a large and diverse region. It is located just above the front of the corpus callosum and curls around in the same way (see figure 2.2). It connects to several parts of the prefrontal cortex in addition to the orbitofrontal cortex, including the dorsolateral and medial areas. The connections of the anterior cingulate to the orbitofrontal cortex are overwhelmingly to limbic areas in the gyrus rectus (Morecraft et al. 1992) (i.e., to Carmichael and Price's medial network). The anterior cingulate's thalamic connections include the mediodorsal and intralaminar nuclei.

Devinsky et al. said that the anterior cingulate cortex "is part of a larger matrix of structures" that "form the rostral limbic system and include the amygdala, periaqueductal gray, ventral striatum, orbitofrontal and anterior insular cortices." This system "assesses the motivational content of internal and external stimuli and regulates context-dependent behaviors" (1995, 279). They divided the anterior cingulate into an affect region, which assesses "motivational content," and a cognitive region, which goes into action during "cognitively demanding information processing," and which also has dense connections to motor areas. The affect portion of the anterior cingulate is located in its anterior, vertical part (in Brodmann areas 25, 33, and rostral area 24) and has connections to the amygdala and autonomic nuclei (see figure 4.3). Devinsky et al. located the cognitive part in the caudal or rearmost portions of Brodmann areas 24 and 32, sometimes called the cingulate motor areas, and in an area known as the nociceptive cortex because of its role in perception of pain.

Temporal Lobe Connections

The temporal lobes specialize in analyzing incoming visual and auditory information, but they also have important mnemonic functions. Two basic types of temporal lobe process communicate with the orbitofrontal cortex—memory-related processes in the interior, medial portion of the temporal lobes and visual processes in the lateral and inferior temporal lobes. The orbitofrontal cortex seems to have separate mnemonic and sensory divisions. "In particular, the medial and orbital areas that are related to the hippocampus are segregated from the lateral and posterior orbital areas that are related to the sensory and premotor areas" (Carmichael and Price 1995a, 655). Carmichael and Price found that the subiculum (see figure 3.1, number 16), an important part of the medial temporal lobe memory system, projects to orbitofrontal areas 13a, 13b, 11m, 14r, and 14c: "These projections from the hippocampal formation are matched by projections to the same cortical areas from the anteromedial thalamic nucleus" (1996, 204). These connections may be important to understanding the role of anterior thalamic damage in Korsakoff's syndrome.

Other parts of the orbitofrontal cortex receive visual input from several of the highest-level visual processing areas in the temporal lobes. Carmichael and Price (1995b) found in a staining study that visual areas throughout much of temporal area TE project to orbitofrontal area 12l. They also discovered that visual areas in the inferior temporal cortex are reciprocally connected with orbitofrontal areas, primarily area 12l. In another staining study (1995a) these authors also found that other temporal lobe areas, the multimodal areas in the superior temporal cortex (see chapter 5), project to orbitofrontal areas 13a and 12o, the common nodes of the orbital and medial networks. One of their staining studies appears to show that these parts of the superior temporal sulcus participate in a network with the orbitofrontal areas they project to, together with areas in the insular cortex, the cortex posterior to the orbitofrontal cortex, which is concealed by the temporal lobe itself. As we will see in chapter 5, these areas become active when we direct our attention toward other people and endeavor to understand their minds.

Dorsolateral Prefrontal Connections

The dorsolateral prefrontal cortex is located at the front and upper side of the brain. One of its specialties is keeping information about an individual's nearby personal space and the objects of interest active long enough for the individual to devise and direct actions at those objects. Hence, as we saw in chapter 3, Goldman-Rakic termed this part of the cortex a "working memory" area. The spatial working memory area receives extensive input from the parietal cortex and apparently is involved in representing an individual's body in its nearby personal space. This area, which centers around Brodmann's area 46 in rhesus monkeys and humans, has been extensively explored (Goldman-Rakic 1987).

According to Goldman-Rakic, multiple working memory systems are adjacent to one another in the prefrontal cortex. Each plays a role in keeping a certain type of information "on line" during relatively short delays. Dorsolateral areas in general are more closely tied to the senses; for instance, the frontal eye fields, which are active in behavioral tasks involving a search of the environment. Another working memory area just below the spatial working memory area specializes in representing objects (Rolls 1999). Goldman-Rakic (1987) suggested that the orbitofrontal cortex might work in tandem with these dorsolateral working memory areas by keeping representations of the emotional valence associated with the representations of objects on line. "The dorsomedial prefrontal region subserves the mnemonic representation of goal states which are the targets of positive and negative emotional states represented, in part, in the ventromedial prefrontal region" (Pandya and Yeterian 2001, 79). It is interesting that the orbital prefrontal network, the one with multimodal input, is interconnected with the dorsolateral prefrontal areas that constitute the working memory area. These connections may be achieving the function of associating what is held in working memory with different reward values contained in the medial orbital network, which were recorded during previous interactions with the relevant objects.

Striatal Connections

The striatum is an oddly shaped structure located beneath the cortex. It consists of the caudate nucleus—which is interconnected with the orbitofrontal cortex—and the putamen. In general, these organs have roles in motor activity, so it is assumed that these connections are a route by which the orbitofrontal cortex can influence and possibly inhibit actions. Zald and Kim (2001), in their comprehensive review of the anatomy and function of the orbitofrontal cortex, pointed out that the functions of the relevant parts of the striatum are not well understood, and that they are sometimes referred to as the "cognitive striatum," because they do not seem to play a direct role in motor execution. Two circuits connect the orbitofrontal cortex and the striatum. The first runs from the orbital cortex to the striatum, then to the globus pallidus, and returns to the orbital cortex by way of projections running through the mediodorsal nucleus of the thalamus (Zald and Kim 2001). A second circuit runs from more lateral portions of the orbital cortex to the head of the caudate nucleus. These connections are segregated, based on their connection to either the medial or the lateral orbitofrontal cortex (Zald and Kim 2001), which seems to again affirm the usefulness of Carmichael and Price's distinction between the two networks: "Thus many of the key structures that interact with the lateral OFC [orbitofrontal cortex] project to the same region of the striatum that is innervated by the lateral OFC, whereas the key structures that inter-act with the medial OFC project to the same region of the caudate and ventral striatum that is innervated by the medial OFC" (Zald and Kim 2001, 46).

Fischer et al. (1995) noted that confabulating patients with ACoA syndrome frequently have damage to the head of the caudate nucleus. Curran et al. tested a patient with a right frontal lobe infarction who showed very high "false alarm" rates on tests of recognition; he frequently claimed to recognize stimuli he had not seen before. The patient's lesion primarily affected motor and premotor areas, but the investigators noted "atrophy of the right caudate nucleus and thalamus" (1997, 1036). Rolls relates its striatal connections to the role of the orbitofrontal cortex in altering behavior that has ceased to be reinforced: "It is via this route, the striatal, that the orbitofrontal cortex may directly influence behavior when the orbitofrontal cortex is decoding reinforcement contingencies in the environment, and is altering behavior in response to altering reinforce-ment contingencies" (1999, 125). Rolls noted further that "neurons which reflect these orbitofrontal neuronal responses are found in the ventral part of the head of the caudate nucleus and the ventral striatum, which receive from the orbitofrontal cortex ... and lesions of the ventral part of the head of the caudate nucleus impair visual discrimination reversal" (1999, 125).

4.4   Sociopathy

Intermittently throughout the twentieth century, sociopaths have also been referred to as "psychopaths," or more benignly, as having "antisocial personality disorder." The orbitofrontal cortex is increasingly coming under investigation as a problem area in sociopathy thanks to two currents in research: the discovery of abnormalities in the orbitofrontal cortices of these individuals, and the discovery that damage to the orbitofrontal cortex in adulthood can make previously normal people begin behaving like soci-opaths, a condition that has come to be known as "acquired sociopathy." In 1975, Blumer and Benson used the term *pseudopsychopathic syndrome* to characterize the lack of social restraint seen in patients with orbitofrontal lesions: "The sociopathic individual recognizes what is right but fails to inhibit socially incorrect behavior, a description which resembles certain behavioral abnormalities that can follow frontal abnormality." This syn-drome is "best characterized by the lack of adult tact and restraints. Such patients may be coarse, irritable, facetious, hyperkinetic, or promiscuous; they often lack social graces and may, on impulse, commit anti-social acts" (1975, 158).

Robert Hare (1993) provided the following list of characteristics of sociopaths: they are glib and superficial as well as egocentric and grandiose; they show a lack of remorse or guilt, as well as a lack of empathy; they are deceitful and manipulative, and they have shallow emotions. Hare also listed criteria directed at the tendency of sociopaths to be social deviants and to get into legal trouble: they are impulsive, have poor behavior controls, and lack a sense of responsibility. Sociopaths are often described as emotionally immature in that their emotions closely match those of children. They can also share confabulators' sometimes outrageous overconfidence in their health and physical condition (see chapter 6). For example, an incarcerated sociopath planned to become a "professional swimmer" when he got out of prison, despite being thirty-eight years old, completely out of shape, and overweight (Hare 1993).

Recall Johnson's reality monitoring theory (section 3.5) that our current thoughts are checked in numerous ways, including a check against various memory resources for contradictions. We also noted the presence of contradictions in the speech of Korsakoff's and ACoA syndromes. Many sociopaths also have a curious inability to notice contradictions in their speech. Hare provides the following example from the same overconfident swimmer: "Asked how he had begun his career in crime, he said, 'It had to do with my mother, the most beautiful person in the world. She was strong, worked hard to take care of four kids. A beautiful person. I started stealing her jewelry when I was in the fifth grade. You know, I never really knew the bitch—we went our separate ways.'" Another man serving time for theft was asked whether he had ever committed a violent offense. He replied, "No but I once had to kill someone" (Hare 1993, 125).

Sociopaths are infamous for their lack of empathy. Hare quoted an imprisoned rapist who found it hard to empathize with his victims: "They are frightened, right? But, you see, I don't really understand it. I've been scared myself, and it wasn't unpleasant." Sociopaths do experience emotions, but these seem to be very shallow. This shallowness seems to bleed over into a shallowness in the emotions they attribute to others. Hare: "Another psychopath in our research said that he did not really understand what others meant by 'fear.' However, 'When I rob a bank,' he said, 'I notice that the teller shakes or becomes tongue-tied. One barfed all over the money. She must have been pretty messed up inside, but I don't know why'" (1993, 44). When this man was asked how he would feel if someone pointed a gun at him, he did not mention bodily sensations. Based on this sort of evidence, Hare concluded that sociopaths lack a conscience because they do not experience the emotions of fear and anxiety strongly enough. Damasio makes the more specific suggestion that the disturbance in sociopaths "sets in at the level of secondary emotions, such as embarrassment and guilt" (2000, 129).

Psychophysiologists discovered in the 1950s that something is awry in the autonomic responses of sociopaths to certain stimuli, especially those that would normally provoke fear or sympathy. As did Damasio, the early researchers primarily used sweating, as measured by skin-conductance changes, as their index of autonomic activity. In an anticipation of the Bechara–Damasio gambling experiment, Lykken (1957) found that sociopaths acquired a conditioned skin-conductance response before an electric shock more slowly than normals. People who had been diagnosed with developmental sociopathy and who had a history of criminal behavior showed the same inability to develop autonomic responses to negative stimuli that Damasio's patient E.V.R. showed (Hare and Quinn 1971). Boys age nine to seventeen with "psychopathic tendencies" also failed the Bechara–Damasio gambling task (Blair et al. 2001). Sociopaths had reduced skin-conductance responses compared with normals during a task designed to be stressful: preparing a list of their flaws and failures, then reading it out loud to the examiner (Raine et al. 2000). In a finding that sheds light on sociopaths' problems with empathy, the overall skin-conductance baseline of sociopaths did not increase as much as that of normal people at the sight of another person apparently being shocked for giving incorrect answers (Dengerink and Bertilson 1975). Hare (1972) found that sociopaths had smaller skin-conductance responses to the insertion of a hypodermic needle than normal persons, and Mathis (1970) found a similar lack of reaction when sociopaths viewed slides of severe facial injuries.

The favorite weapon of the sociopath is language—words used smoothly, articulately, and with complete confidence. Recall the discussion in chapter 1 about whether confabulators are lying. Most researchers agree that they are not. In sociopaths, however, we seem to move farther down the continuum to clear cases of intentional lying, yet not all the way. Sociopaths are capable of lying skillfully, but they come close to confabulators in that they sometimes seem to believe what they are saying. Hare says: "Many observers get the impression that psychopaths are unaware that they're lying; it is as if the words take on a life of their own, unfettered by the speaker's knowledge that the observer is aware of the facts. The psychopath's indifference to being identified as a liar is truly extraordinary; it causes the listener to wonder about the speaker's sanity" (1993, 48).

Damage to prefrontal lobes in infancy produces a condition very similar to what is known as "developmental antisocial personality" (Anderson et al. 1999). Two people who sustained ventral prefrontal cortical lesions before the age of sixteen months and who were twenty and twenty-three years old at the time of study, were described as lying chronically, "often without apparent motive," were unable to experience empathy, and were constantly in trouble with one authority or another. Both had extensive frontal lesions; in one patient, exclusively in the right hemisphere (Anderson et al. 1999).

4.5 Lying and the Skin-Conductance Response

Most people would claim that they have some ability to discern whether or not they are being lied to, but this may itself qualify as a case of confabulation, since it is much more difficult to detect lying than we imagine. Numerous psychological studies show that normal people have no better than random success at detecting lying (Ekman 1992). The most popular folk criterion for detecting lying is of course nervousness on the part of the liar. Nervousness certainly *can* indicate lying, and one might say that it is what the polygraph measures, but nervousness may have other causes. Conversely, a lack of nervousness indicates nothing. This may be one of those cases in which we unconsciously make a fallacious inference. From the already flawed hypothetical, if a person is nervous, he is lying, we mistakenly conclude that if a person is not nervous, he is not lying (compare: if a person is an uncle, he is a man, therefore, if a person is not an uncle, he is not a man). Cool, confident-appearing sociopaths capitalize on this poor reasoning to deceive their victims.

The polygraph, or lie detector test, consists traditionally of three measures: skin conductance, breathing rate, and blood pressure. Of these three, polygraph operators rely primarily on skin-conductance changes, which have been shown to be the most reliable indicator of deception (Waid and Orne 1981). The accuracy of these tests has long been a sore point among supporters and critics of their use, but it is far enough below 100 percent that most states ban them from being used as evidence in trials. The attempt to use experiments to settle the issue is muddied by a huge technical problem: When criminals lie, there is often a great deal at stake. They are typically tense and anxious when being tested, and this may be essential to producing responses detectable by the polygraph operator. How can one recreate such a situation in a laboratory using standard college undergraduates as subjects (without being sued)?

Skin-conductance responses, historically known as galvanic skin responses, are easy to measure yet allow a high-resolution measure of activity of the autonomic nervous system. They are produced by activity of the eccrine sweat glands, located on the palms and soles of the feet (as opposed to the bodywide apocrine sweat gland system, which functions primarily to regulate body temperature). Along with dilation of the pupils and decreased motility of the gastrointestinal tract, palmar sweating produced by the eccrine system is part of the sympathetic (fight or flight) branch of the autonomic nervous system. Because of their connections to high-level brain systems, the eccrine sweat glands provide a window onto certain types of activities having to do with emotions.

No one has ever tested confabulating patients to determine whether they produce any autonomic indices of lying. Evidence leads us to think that they would pass a standard polygraph test, however, owing to a hyporesponsive autonomic system. Patients who denied their illness (see chapter 6) and showed a lack of emotion also failed to produce SCRs (Heilman et al. 1978). Techniques may exist that can show that confabulators are not using the processes normal people use when they make correct memory reports, such as the studies of Schacter et al. (1996) described in chapter 3 that detected different brain activities associated with true and false memory reports. But to correctly claim that a person is genuinely lying, the intention to deceive must also be detected.

As we noted earlier, the general trend is toward finding that sociopaths are hyporesponsive in SCR tests. Fowles states that "the phenomenon of electrodermal hyporeactivity in psychopathy is one of the most reliable psychophysiological correlates of psychopathy" (2000, 177). This sort of hypoarousal was correlated with "uninhibited" behavior (Jones 1950), and consistent with this, Hare (1993) reported that sociopaths have a need for excitement. This hyporesponsiveness may also help explain why sociopaths are able to lie effectively; they are not nervous when doing so. If you or I were caught lying and publicly confronted with it, we would experience deep embarrassment and shame. But when sociopaths are caught, it does not seem to even slow them down (Hare 1993). Their typical response is to come up with another lie. In one study in which subjects attempted to deceive experimenters, their behaviors such as eye movements and facial expressions decreased when they were actually lying (Pennebaker and Chew 1985). This behavioral inhibition coincided with increases in skin-conductance level, a piece of information that may help link what is known about the lying response to what is known about the role of the orbitofrontal lobes, both in inhibition and in producing SCRs.

Anatomy of the Skin-Conductance Response

Several different cortical areas are capable of initiating SCRs. These may be subdivided into two types: those that produce SCRs related to bodily activity, such as hand clenching or deep breathing, and those that produce SCRs related to psychological activity, such as the sight of a significant person or of a gruesome murder. There are motor areas that initiate skin-conductance responses; one of the standard ways to begin an SCR test is to have the subject tightly clench a fist (Sequeira and Roy 1993). Our interest is in the psychological SCRs, which have an intriguing lateral component. There is a strong trend toward finding that right hemisphere damage reduces skin-conductance activity (there is more on this in chapter 7). Studies also show that left hemisphere damage can increase skin-conductance activity, perhaps owing to a Jacksonian releasing effect. Many of the brain areas we have looked at, or will look at because they are relevant to the problem of confabulation, are also areas involved in production of SCRs.

Anterior Cingulate   Electrical stimulation of the anterior cingulate cortex in animals produces skin-conductance responses (Neafsey 1990). Critchley et al. (2001) found that activity in the anterior cingulate was correlated with skin-conductance responses and with the amount of uncertainty their subjects experienced. These researchers were able to vary the amount of uncertainty produced in subjects during each trial using the following technique. The subjects saw one of ten cards, valued at one through ten. Their task was to determine whether the next card would be higher or lower than the first one. If the original card is a one, chances are quite high the next card will be higher, so there is little uncertainty in making this choice. However, if the first card is a five or a six, uncertainty is greater since the predicted success of the choice is near fifty percent. The authors suggest that "the same subregion of anterior cingulate represents both cognitive uncertainty and physiological arousal" (2001, 543). A connection also may exist between confabulation and other symptoms of frontal damage—the ability to make estimates (Beeckmans et al. 1998) and the ability to make informed guesses. Guessing activates the ventromedial prefrontal cortex (Elliot et al. 1999). Making a guess requires weighing several different candidates and selecting the most probable one. In contrast, confabulators seize on the first explanation that pops into their minds and appear to be unable to evaluate its plausibility.

Orbitomedial Cortex   I noted earlier that the orbitofrontal cortex also has projections to the lateral hypothalamic nucleus, both by way of the amygdala and directly. Critchley et al. (2000, 2001) found activity in the right orbitofrontal cortex (Brodmann's areas 11/47; figure 4.3) during a gambling task, just as Damasio's research would predict. Fowles (1980) theorized that inhibiting behaviors itself produces electrodermal activity, something the Pennebaker and Chew (1985) experiment described earlier seems to affirm. Also, patients with diffuse damage to the orbital and ventromedial cortex that is due to a closed head injury failed to show a normal SCR to negative facial expressions (Hopkins et al. 2002). Notice how the orbital and medial areas indicated in figure 4.5 correspond well with Carmichael and Price's medial network (see figure 4.4).

Right Inferior Parietal Cortex   Several studies have established that the right inferior parietal cortex is capable of generating skin-conductance

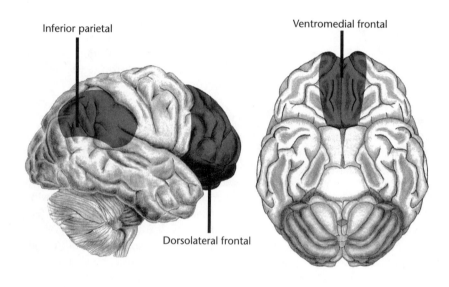

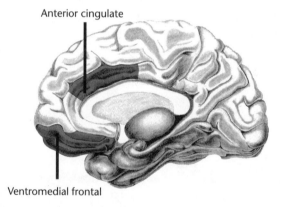

Figure 4.5
Diagram of cortical lesion sites that can affect the skin-conductance response.
(Based on a figure from Tranel 2000.)

responses. This response may travel from the inferior parietal area to the basal forebrain (see Mesulam et al. 1977). We examine this area in greater depth in chapter 6 when we look at patients who deny they are ill, many of whom have right inferior parietal damage.

Temporal Lobes–Amygdala   Through its connections with the lateral hypothalamus, the amygdala has what is thought to be an excitatory role in producing autonomic responses, such as pupil dilation, decreased gastric motility, and palmar sweating. Raine et al. (1991) found a correlation between the number of SCRs to an orienting stimulus and the size of the left temporal-amygdala region. Direct electrical stimulation of the amygdala produces strong SCRs in humans (Mangina and Beuzeron-Mangina 1996), and ablation of the amygdala in rhesus monkeys decreases the size of SCRs to tones (Bagshaw et al. 1965).

One route by which a psychological SCR may be produced by visual perception begins in areas of the temporal lobes responsible for analysis of an object, then progresses to the lateral nucleus of the amygdala, where it travels to the basal nucleus (Amaral et al. 1992; Halgren 1992). A circuit is completed when the amygdaloid basal nucleus returns projections to the same areas of the temporal lobes. Also though, the basal nucleus projects to the central nucleus, which is considered the output center of the amygdala. Through its connections to the hypothalamus and directly to the nucleus of the solitary tract, the central nucleus can initiate sympathetic activity (Amaral et al. 1992). However, the role of the amygdala in generating SCRs has been questioned, because patients with bilateral amygdala destruction generated normal SCRs (Tranel and Damasio 1986; Tranel and Hyman 1990; see also Lee et al. 1988). The role of the hypothalamus in producing SCRs is more certain. A review of the role of the hypothalamus and limbic system in electrodermal activity states, "taken together, data from the three hypothalamic regions indicate that the anterior, medial, and posterior hypothalamus exert an excitatory influence on [electrodermal activity]" (Sequeira and Roy 1993, 95).

4.6   Obsessive-Compulsive Disorder as a Mirror-Image Syndrome

If damaged orbitofrontal brain areas cause confabulation, what happens in the mirror-image case, when orbitofrontal areas hyperfunction? Accounts of the neural substrate of obsessive-compulsive disorder trace it to a hyperfunctioning orbitofrontal cortex (Schwartz 1998; Saxena et al. 1998). Patients with OCD show increased orbitofrontal blood flow when exposed to stimuli that provoke their obsessions (see Zald and Kim 2001). Some symptoms of OCD might be describable as pathological doubt, for instance, that one's hands are clean or that the doors are locked. A review of OCD treatments states, "pathological doubt is one of the central manifestations of this illness. The person goes to the door, shuts it, locks it, feels that it is locked, knows that it is locked, turns around, and walks away. All of a sudden, he or she feels that it is absolutely necessary to go back and check. It appears clinically that the memory of the action of locking the door is insufficient" (Mindus et al. 2001, 241). In this respect, patients with OCD might be seen as the mirror image of confabulating patients, who are unable to doubt. The person with OCD seems to be trying to raise her epistemic standards, her standards of certainty, to absurdly high levels.

Sociopaths' need for excitement can cause them to engage in extremely risky behaviors. "Abnormal risk assessment" is also involved in OCD, but in the other direction. "Patients may incessantly ruminate about the risk of being contaminated with, say HIV, and shun public bathrooms, although they admit knowing that the risk is negligible" (Mindus et al. 2001, 241). Of great interest for our purposes, OCD was connected both to the orbitofrontal cortex and to acquired sociopathy: "OFC hyperactivity may relate to the very processes that appear so deficient in the pseudo-psychopathic condition. For instance, whereas the pseudopsychopathic individual participates in risky behavior despite knowledge of the risks, OCD patients appear excessively concerned with such risks" (Zald and Kim 2001, 60). The two conditions are also mirror images in terms of their effects on the person's social behavior: "The pseudopsychopathic individual shows a lack of concern and an irresponsibility toward others, but OCD patients often show excessive concern with how their actions will affect others.... Similarly, feelings of guilt, shame, anxiety, and concern for social norms, which seem so lacking in pseudopsychopathic persons, appear accentuated in OCD patients" (Zald and Kim 2001, 60).

These authors offer an illuminating speculation as to the exact role of the orbitofrontal cortex in obsessive-compulsive disorder: "OCD patients frequently perceive that they performed previous responses inadequately. If the error coding of OFC cells were hypersensitive, such that responses consistently were coded as errors, individuals might repeatedly experience their responses as inadequate. In extreme cases, this could lead to repetitive attempts to perform acts 'just right'" (Zald and Kim 2001, 60).

A woman who underwent lobotomy, an officer in the Salvation Army, obsessed over the idea that she had committed sins against the Holy Ghost (Rylander 1948). The operation seemed to work: "After the dressing had been taken off, I asked her 'How are you now? What about the Holy Ghost?' Smiling, she answered, 'Oh, the Holy Ghost; there is no Holy Ghost'" (1948, 702). What is most debilitating about OCD is its intrusiveness, the way that the obsessive thought is capable of making the person stop everything and attend to it, go wash his hands, or whatever it demands. "The inability of OCD patients to inhibit intrusive thoughts and images could reflect a hyperactive working memory process in which internally generated representations (expectancies) are maintained indefinitely in a state of moment-to-moment awareness. In the extreme case, this information might become locked 'on-line,' despite repeated attempts to eliminate the representation" (Zald and Kim 2001, 59).

Neurosurgeons have devised a more focused operation for people with debilitating OCD, called subcaudate tractotomy, which is designed to destroy the white matter tracts connecting the orbitofrontal cortex to the thalamus and other structures, or in a variation, to destroy the portion of the orbitofrontal cortex where these white matter tracts enter. This operation has proved to be reasonably successful in relieving the symptoms of severe OCD. In a follow-up study of 134 patients, it produced good results in fifty percent of the patients who had obsessive neuroses (Goktepe et al. 1975). In addition, "transient disinhibition" was common after the operation, as we might expect. Closer to home, Kartsounis et al. (1991) noted that a large proportion of patients who undergo this procedure have a marked tendency to confabulate on recall tasks. They also performed poorly on a test of their ability to estimate.

4.7   Conclusion

There appears to be a continuum, with confabulators and sociopaths at one end and people with OCD at the other. Confabulators, and possibly sociopaths, have damage to circuits passing through the orbitofrontal cortex, whereas these same circuits are hyperactive in OCD. In chapter 9 I will place other types, including self-deceived normal people, on this continuum.

It is significant that many of the areas capable of generating psychological skin-conductance responses are also implicated in confabulation: the orbitomedial cortex, anterior cingulate, temporal lobes (see chapter 5), and inferior parietal cortex (see chapter 6). Several clues here point to the tantalizing possibility that the skin-conductance response used to detect lying originates in the orbitofrontal cortex. A second plausible candidate for production of the lying response is the anterior cingulate cortex. Lying involves inhibiting one response and issuing another, and the anterior cingulate is active in tasks in which this sort of thing is required (see Posner and Raichle 1994).

Lying as well as repeatedly making false claims is socially inappropriate behavior, and lying is one of the most antisocial things the sociopath does. Many confabulators are socially disinhibited. Sociopaths engage in antisocial behavior, perhaps because of a lack of inhibition. How does social inhibition work? What is it in us that keeps us from doing what sociopaths do? A related question is what in us allows us to express empathy toward other people, whereas sociopaths do not. These are some of the questions the next chapter addresses when we look at confabulators who make shocking claims about the identities of people close to them: those with Capgras' syndrome.

5 Mind Reading and Misidentification

He's a nice man, doctor, but he's not my father.
—Patient with Capgras' syndrome (Hirstein and Ramachandran 1997)

5.1 Knowledge of Others' Minds

The patient with Capgras' syndrome believes that a person close to him or her, typically a spouse or family member, is actually an impostor. These patients seem to confabulate, particularly when they explain how and why people are not who they appear to be. However, it has proved difficult to integrate what is known about this disease with what is known about the other confabulation syndromes, and so far few attempts have been made to provide a conception of confabulation capable of including both the memory syndromes we examined in chapter 3 and misidentification syndromes such as Capgras'. My main goal in this chapter is to examine Capgras' syndrome and the other misidentification syndromes in which patients create confabulations about the identities of people close to them, and integrate what is known about them with what we have already learned about the neural basis of confabulation.

Confabulatory patients have a problem that has received little attention up to now, but I contend that it is an important clue to their condition: They have a gross insensitivity to their hearers. "They maintain their fictitious stories despite cues, corrections, and the obvious disbelief of their auditors" (Weinstein 1971, 442). A normal person would respond to obvious evidence that his audience does not believe him. He might become angry or feel nervous, depending on the circumstances, but almost anyone would try harder to convince his listeners. The normal person would at least acknowledge his listeners' incredulity by saying something like, "I know this sounds odd, but." An important connection between memory confabulators and those with Capgras' syndrome comes, I maintain, by way of this feature: a mind-reading problem.

In chapter 3 we explored forms of confabulation that accompany two memory disorders, Korsakoff's syndrome and ACoA aneurysm. We saw there that damage to certain orbitofrontal areas or their vital connections affects a patient's ability to judge whether what seems to him to be a memory really is an accurate representation of reality. In chapter 4 I argued that a closer look at the orbitofrontal cortex and other syndromes caused by its damage uncovers a family of syndromes to which confabulation belongs, including sociopathy and disinhibition. Orbitofrontal damage has been reported in sociopaths, and many sociopaths seem to alternate between being confabulatory and confidently lying. Lying by normal people, on the other hand, may sometimes be accompanied by autonomic reactions, which both confabulators and sociopaths appear to lack. Such reactions may serve an inhibitory function. They keep us from making false claims just as much as they keep us from engaging in other risky behaviors.

In the view I propose, confabulating is a behavior that results from disinhibition just as much as the behavior of disinhibited patients who make socially inappropriate remarks. Confabulations emerge from a cognitive system that would have inhibited them, had it been intact. It may be true that confabulation involves a prior problem, in that the thoughts that the patient's cognitive system generates might not have been generated if there had not been some damage to that system. However, if the same ideas had occurred in *our* minds, we would not have offered them up for public consumption as explicit claims. Just as confabulators do not realize the ill-groundedness of their claims, they fail to realize that their listeners *do* find their claims ill-grounded. Why don't they see that their claims will not be believed? They certainly do not seem worried or concerned that they are not being believed, as evidenced by their calm outward demeanor and hyporesponsive SCRs. Recall Hare's (1993) similar remark about the indifference of sociopaths to being identified as liars.

Problems in social interaction were noted in people with both Korsakoff's syndrome and ACoA aneurysm. Similarly, in our examination of the symptoms of orbitofrontal damage, we frequently came across mention of problems the patients had in their social interactions with others. Many of them seemed to have lost their ability to consider the feelings of others. Sociopaths do not seem to care much about the feelings of other people either, and they seem not to model the mental states of the people they interact with very thoroughly. They may know the basic principles of how to con someone, e.g., flatter the person, add convincing details, but this can easily go wrong, and this is one area in which they are often insensitive. Their failure to understand their listeners seems to be based on their inability both to know what their feelings are like and to spontaneously represent them as occurring in other people. Sociopaths may thus be poor at certain types of emotional understanding.

One can of course read facial expressions without having much of a theory of mind; one can act as a behaviorist might in dealing with other people, and this may be what sociopaths do. However, it is not hard to fool the sociopath into thinking that you believe his line. Who has not experienced the near-sociopathic salesman, very slick and convincing in his pitch, perhaps because he either believes it himself or can suspend disbelief while he is making it. These types seem poor at differentiating interest in what they are saying from mere tolerance. Recall Hare's (1993) intriguing remark that many observers have the impression that psychopaths are unaware that they are lying. The success sociopaths have at lying may come more from their confidence than from an ability to understand the minds of others. Lying truly effectively requires knowledge of the other person's mind; this is a variety of mind reading that targets beliefs rather than emotions. The liar has to know that the mind of his victim is in the proper state to be deceived (more on this in chapter 9). He knows he needs to avoid skeptical or sharp-witted people, but even these people can be lied to successfully if one is able to play to their weaknesses. Sociopaths typically leave them alone, however; they are experts at picking their marks from the ranks of the gullible. If sociopaths are bad at mind reading, this might explain many criminal partnerships, as well as their frequent dissolution; sociopaths can con each other.

Our awareness at some level that confabulators are insensitive to the minds of their listeners may be another factor in our reluctance to consider them liars. There are thus two reasons for the claim that confabulators are not lying: They have the beliefs expressed by their confabulations; that is, they are frequently reported to act according to them; and they are insensitive to the incredulity of their listeners. This mind-reading hypothesis goes against the idea implicit in Bonhoeffer's phrase "confabulations of embarrassment." This notion has not received support. "It is not consistent with clinical observations to call these replies 'confabulations of embarrassment' as the patient shows not the least embarrassment when confronted with the true facts of his present situation" (Whitlock 1981, 216). Hare is clear that sociopaths do not suffer embarrassment either. Embarrassment is a social emotion (Panskepp 1998) and can be a powerful motivating force, although this varies greatly from person to person. When the thought of engaging in a certain action occurs, the normal cognitive system then attempts to model how events will progress after the act is committed, including how other people might react. Another part of this modeling process is the brain's ability to attach a value to the envisaged representations, exactly the sort of thing the orbitofrontal cortex specializes in according to Rolls (see section 4.2).

The fact that orbitofrontal damage can produce disinhibition may be important here. It may be that patients with this damage engage in socially inappropriate behavior because they are not imagining how their actions will look to those around them. This ability, which sociopaths and confabulators also seem to lack, is one of a set we use to understand others that have been called mind-reading abilities (Baron-Cohen 1995), or "theory of mind" (Premack and Woodruff 1978). Mind reading is broader than another term sometimes used in this context, simulation, since there may be ways we read minds without simulating or modeling them. I might draw inferences about your mental state by reasoning from your behavior, without simulating you in any way. Mind reading involves representing the other person's mental states, but this includes merely having a belief that, for example, he is angry, without simulating his anger.

At the extreme end of malfunction of the mind-reading systems is autism. People with autism characteristically have trouble with several mind-reading tasks. One connection among mind reading, social emotions, and autism is idea that children have to look at adults or other role models to gain information about appropriate emotional responses to events. Autistic children do not do this. For example, when they were exposed to a person in (simulated) distress, they showed a "lack of social referencing; they did not tend to look toward an adult in the presence of an ambiguous and unfamiliar stimulus" (Bacon et al. 1998, 129). High-functioning autistic children are infamous for asking socially inappropriate questions. In general, autistic people do not experience normal social emotions.

In chapters 3 and 4 we saw that the orbitofrontal cortex is a site of interest in the study of confabulation, and that other syndromes that involve orbitofrontal damage, such as sociopathy and disinhibition, share important features with confabulation. The orbitofrontal cortex connects to Capgras' syndrome in two ways. First, patients with this syndrome often have orbitofrontal damage. We may be able to understand the relationship between Capgras' syndrome and the confabulation syndromes we have seen thus far by tracing the inputs to the orbitofrontal cortex back one step to the temporal lobes. Second, the orbitofrontal cortex is part of a mind-reading system, that I believe malfunctions in the misidentification syndromes. In this hypothesis mind reading is the link between the orbitofrontal cortex and Capgras' syndrome. More generally, I suggest that some of the checking processes damaged in confabulating patients are implemented by mind-reading circuits. This can help us begin to understand the connections among social disinhibition, sociopathy, and confabulation.

5.2   Mind-Reading Systems

We are still in the early stages of learning how we understand each other's minds, but several breakthroughs have occurred. Primarily from lesion, cell-recording, and imaging studies, we are discovering which parts of the brain are doing the heavy lifting. First, it must be clear that theory of mind or mind reading involves several different, possibly even dissociable types of tasks. We seem to have both first-person (or "internal") and third-person (or "external") ways of accomplishing each one, such as the following:

1. Understanding the perceptions of others   This can involve understanding a person in an environment as he relates to objects of interest and predicting his actions in that environment, as well as simulating the perceptual point of view of a person in an environment. This would include the interesting phenomenon known as *joint attention*: You and I attend to an object, and each of us is aware that the other is also attending to it.

2. Understanding the actions of others   One way to do this is by simulation, pretending at some level that we ourselves are simultaneously performing the same action. We can also use the backward inference described earlier to infer mental states from behavior.

3. Understanding the beliefs of others   This is a subcase of understanding and predicting a person's mental states. It can involve simulation of the beliefs of others, especially false beliefs.

4. Understanding the emotions of others   We often simulate the emotions of others, a process that occurs automatically in normal people but possibly not in sociopaths. When I simulate someone, I expect a certain pattern of emotions in that person. I know which emotions generally follow which; for example, frustration is typically followed by anger.

Leslie Brothers (1995) distinguished cold and hot approaches to the study of mind reading. Cold accounts focus primarily on the way that we understand the attention, true beliefs, and false beliefs of our targets. A hot approach, on the other hand, focuses on the way that we model the emotions, or "evaluative attitudes" of others. "Evaluative signals, in combination with the direction of attention as indicated, for example, by gaze, are used by the observer to create descriptions of intention and disposition. Normal children use 'social referencing' in ambiguous situations to discern the evaluative attitudes of caregivers toward novel stimuli" (Brothers 1995, 1113). While she emphasized the importance of autonomic reactions, Brothers differentiated her approach from that of Damasio (see section 4.2). Damasio's account is strictly dichotomous in that somatic messages signal either reward or punishment. Alternatively, Brothers postulated that many different social situations are linked with many different, discriminable patterns of autonomic responses.

Another way to understand the different functions that comprise the human skill for mind reading is to type them by tests that have been devised by cognitive neuropsychologists. In a first-order false belief task, the subject knows that the target person has a false belief and must use this to predict his behavior. One way to implement this is by storytelling in which pictures must be arranged to tell a coherent story, and doing this successfully requires knowledge of the character's false belief. In second-order false belief tasks a person has a false belief about the beliefs of another person. Understanding of what philosophers and linguists call conversational implicature also requires knowledge of others' minds. Said in the right context, the phrase, "It's very cold in here," should also be taken to communicate a request to close the window. Another way to test mind reading is to see whether subjects understand certain jokes that require knowledge of the mental states of the characters in the joke.

The brain also has special tricks it uses to understand people's actions. Giacomo Rizzolatti and his colleagues were recording the firing rates of individual neurons in the frontal motor cortex (ventral area 6) of rhesus monkeys as they performed hand actions, such as taking an almond out of a cup (Rizzolatti et al. 1996). When one of the researchers returned from lunch, he licked his ice cream cone in the monkey's view, causing the neurons to fire, to the researchers' great surprise.

These investigators went on to discover these "mirror neurons" in several brain areas. Mirror neurons are most effectively activated by the sight of simple hand and arm actions, such as picking up an object. They receive both visual input from a ventral visual stream, and somatosensory input, and they are found in both motor and somatosensory areas. Apparently the brain uses its own body representations to allow us understand the actions of others. As we will see, other areas of the brain respond to the sight of a moving face, and also have connections to areas known to be involved in emotions, including the orbitofrontal cortex. These areas may be crucial to our ability to understand other's emotions.

Brothers (1990) singled out certain cortical areas as important for mind reading, including the orbitofrontal cortex and the superior temporal sulcus of the temporal lobe (figure 5.1). According to Baron-Cohen, a circuit connecting the orbitofrontal cortex, the superior temporal sulcus, and the amygdala is "the seat of the mindreading system" (1995, 93). Let us examine the evidence that these and other areas have a role in mind reading.

Orbitofrontal Cortex

Several studies point to a role for the orbitofrontal cortex in mind-reading tasks. Patients with bilateral orbitofrontal lesions performed similarly to individuals with high-functioning autism (sometimes called Asperger's syndrome) in a social reasoning task in which they had to recognize when someone had committed a faux pas (Stone et al. 1998). In one example, someone mistakenly revealed to her friend that a surprise party was planned for her. In another, a woman told her friend that she never liked a certain crystal bowl after it broke, forgetting that the friend gave it to her (Stone et al. 1998). Profound theory of mind deficits were observed in patients with ventromedial frontal damage, who no one had suspected had such deficits (Gregory et al. 2002). These patients performed poorly in detecting faux pas and in a test devised by Baron-Cohen et al. (1997) called the Reading the Mind in the Eyes Test, in which the patient looks at a

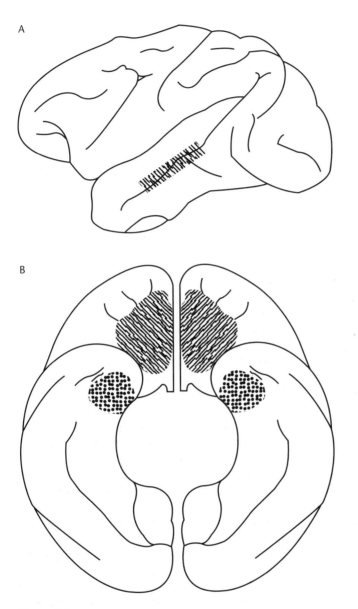

Figure 5.1
Areas important for mind reading, according to Brothers. (*A*) The superior temporal sulcus. (*B*) The orbitofrontal cortex and amygdala. (Based on a figure from
Brothers, L. 1990. The social brain: A project for integrating primate behavior
and neurophysiology in a new domain. *Concepts in Neuroscience* 1: 27–51.)

photograph of a person and must decide whether the person is, for instance, ignoring or noticing the viewer.

"Empathic processing" was compared in four groups of patients with frontal damage—left dorsolateral, right dorsolateral, mesial, and orbitofrontal. Those with orbitofrontal damage were most impaired in their empathic abilities (Grattan and Eslinger 1989). Patients with bilateral, particularly right, orbital/medial damage performed poorly in detecting that another person was trying to deceive them (Stuss et al. 2001). In their deception condition, a confederate of the experimenter points at the wrong cup in a task where the subject has to guess which cup a quarter is under. The confederate has a genuine motive to deceive, since he gets to keep the quarter if the subject guesses incorrectly.

Imaging studies also reveal activity in orbitofrontal areas during tasks such as recognizing emotions and remembering faces. Activity was noted in the right inferior frontal cortex (area 47; figure 4.3) in subjects who were asked to describe the emotions present in faces shown to them (Nakamura et al. 1999). A group of patients with ventral frontal lobe damage and socially inappropriate behavior were also impaired in their ability to identify emotions present in faces and voices (Hornak et al. 1996). The investigators attributed both problems to orbitofrontal damage, and suggested that the social problems are caused by disinhibition. They related their findings to both face processing and social behavior by suggesting that information about faces reaches the orbitofrontal cortex, "where it could be used to provide reinforcing signals used when social and emotional behavior must be altered" (Hornak et al. 1996, 248). In another study, activity of neurons in the orbitofrontal cortex responded to face gesture and movement. "The significance of the neurons is likely to be related to the fact that faces convey information that is important in social reinforcement, both by conveying face expression ... which can indicate reinforcement, and by encoding information about which individual is present, also important in evaluating and utilizing reinforcing inputs in social situations" (Rolls 1999, 124). A review of the functions of the orbitofrontal cortex affirmed this: "The emotional reactions of other people that occur in response to one's own behavior often act as primary or secondary reinforcers and provide important information about when one needs to modify one's behavior. It is not difficult to see how the lack of such abilities could lead to socially inappropriate behavior, especially in someone with diminished inhibitory control" (Zald and Kim 2001, 53).

Acts of confabulation fit this template. Providing false or implausible information to someone without good reason is considered socially inappropriate, but confabulators are unable to notice they are doing this and correct themselves. The entire phenomenon cannot ultimately be due to a failure to gauge the reactions of one's listeners since by then the act of confabulation has already taken place. However, it may explain why the original confabulation is not subsequently corrected, based on the reactions of one's listeners. This inability may play an important role in making a person generally confabulatory, in that he or she persists in tossing out ill-grounded claims.

Amygdala

In chapter 3 we saw that the amygdala is part of a circuit involving the orbitofrontal cortex and thalamic mediodorsal nucleus, damage to which is strongly implicated in causing confabulation. In chapter 4 we examined some of the roles of the amygdala in emotion and in generating autonomic activity. Recently, several groups of researchers have argued for an important role of the amygdala in mind-reading tasks. The amygdala theory of autism (Baron-Cohen et al. 2000) continues an earlier line of inquiry (Brothers 1989; Brothers and Ring 1992). Several pieces of evidence indicate that amygdala dysfunction may play a role in autism. When the amygdalas and nearby temporal cortex of newborn rhesus monkeys were ablated, as the animals grew up they exhibited several of the characteristic features of autism, including active withdrawal from other monkeys and motor stereotypies, such as constantly spinning or somersaulting; and they also had blank, expressionless faces (Bachevalier 1994, 1996). Histopathological examination of autistic brains revealed abnormalities in the amygdala, as well as other parts of the limbic system (Bauman and Kemper 1994). In addition, the amygdala has neurons sensitive to gaze direction, and autistic children have difficulty interpreting gaze information (Baron-Cohen 1995; Baron-Cohen et al. 1997).

The high rate of epilepsy in autistic children may also indicate a role for amygdala pathology in autism. Using magnetoencephalography (MEG), Lewine et al. (1999) found epileptiform activity in eighty-two percent of fifty autistic children. Of the autistic children known to have seizures, seventy percent have a seizure focus in the temporal lobes. Seizures in the temporal lobes are likely to involve the amygdala—it is one of the most kindling-prone organs in the brain (Goddard et al. 1969). Kindling, or seizure activity, tends to spread; The orbitofrontal lobe's "widespread connections with the amygdala, hippocampus, temporal pole, insula, cingulate cortex, and the parahippocampal gyrus may also explain why orbitofrontal cortex is so frequently involved in the origin and spread of temporal lobe epilepsy" (Morecraft et al. 1992). In 1997, hundreds of Japanese children watching a cartoon had seizures, apparently caused by the flashing eyes of one of the cartoon characters, a phenomenon that highlights the connection between sensitivity to gaze and seizures.

As noted in chapter 4, epileptologists sometimes electrically stimulate the brains of their patients prior to surgery to determine the functions of areas they are considering removing to prevent seizures. Merely applying the right voltage to the cortex can trick the brain into activating its circuits (Penfield 1975). These probes can produce a fascinating variety of experiences, many of which seem to involve social emotions. When epileptologist Pierre Gloor stimulated the amygdala, it produced the predictable autonomic reactions and experiences of fear, but it also reliably reproduced recall of episodes from autobiographical memory. These episodes are "most often set in ... a social context.... They frequently bear some relationship with familiar situations or situations with an affective meaning. They frequently touch on some aspect of [the patient's] relationship with other people, either known specifically to him or not." When the left amygdala of a patient was stimulated, it produced in the man "a feeling 'as if I were not belonging here,' which he likened to being at a party and not being welcome" (Gloor 1986, 164). Stimulating the right hippocampus of one patient produced "anxiety and guilt, 'like you are demanding to hand in a report that was due 2 weeks ago ... as if I were guilty of some form of tardiness'" (1986, 164).

Superior Temporal Sulcus

As they progress from the primary visual areas at the back of the brain forward toward the prefrontal lobes, the temporal lobes' ventral visual streams branch and head toward two different areas. Some have connections to working memory areas in the lateral prefrontal cortex (Goldman-Rakic 1987), and others have connections to our area of interest—the orbitofrontal cortex. Perrett et al. (1992) found cells in the superior temporal sulcus that respond to the sight of faces and also to the sight of goal-directed actions (figure 5.2; Carey et al. 1997). This brain area receives visual input from a parietal area involved in the "where" stream, PG, and is connected directly to orbitofrontal areas 13a and 12o (Carmichael and Price 1995a; see also section 4.3). According to one theory, this area specializes in analyzing facial movements, especially those involved in expressing emotions (Haxby et al. 2000). We will look in greater detail at this theory and at this brain area in our discussion of the misidentification syndromes.

Inferior Parietal Cortex

Damage to the right inferior parietal cortex can produce a deficit in recognizing facial expressions (Adolphs et al. 1996). As we will see in chapter 6, this is a high-level somatosensory and visual area. The following discussion suggests why we might see activity in a somatosensory area when subjects are discerning facial expressions of emotion: "In addition to retrieval of visual and somatic records, the difficulty of the particular task we used may make it necessary for subjects to construct a somatosensory representa-

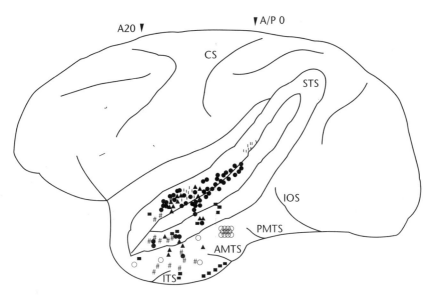

Figure 5.2
Face-responsive cells in the temporal lobes. Each symbol type represents face-sensitive cells found by a different study. (From Perrett, D. I., Hietanen, J. K., Oram, M. W. and Benson, P. J. 1992. Organization and function of cells responsive to faces in the temporal cortex. *Philosophical Transactions of the Royal Society of London*, B 335: 23–30, with permission.)

tion by internal *simulation*. Such an internally constructed somatic activity pattern—which could be either covert or overt—would simulate components of the emotional state depicted by the target face. In essence, the brain would reconstruct a state akin to that of the person shown in the stimulus" (Adolphs et al. 2001, 101).

Staining studies also appear to show that a portion of the posterior cingulate cortex (see figure 2.2) is involved in this system of areas (Vogt and Pandya 1987); an injection to the posterior cingulate in a rhesus monkey labeled the medial orbitofrontal cortex, the superior temporal sulcus, and the parietal cortex. The orbitofrontal cortex also has important connections to motor and somatosensory areas. Ventral motor area 6vb projects "heavily" to orbitofrontal area 13l (lateral) and area 6va projects to area 12m (Carmichael and Price 1995b). Ventral area 6 is where Rizzolatti and colleagues first found mirror neurons; these orbital connections may be the way in which these neurons are integrated into the larger task of mind reading. Several somatosensory areas also connect to orbitofrontal area 12m.

Lateralization of Mind-Reading Functions

Several studies point to a special role for the right hemisphere in mind reading. Patients with right hemisphere damage had trouble with a basic theory of mind task, a first-order false belief task such as the following: "Sam thinks his puppy is in the garage, but the puppy instead is really in the kitchen. Where will he look for the puppy?" (Siegal et al. 1996). A comparison group of patients with left hemisphere damage performed at normal levels. Patients with right hemisphere damage also have trouble distinguishing lies from jokes (Winner et al. 1998). Examples were set up so that one had to know the state of mind of the speaker in order to discern whether he was lying or joking. In one example, a man's boss sees him at a hockey game after the man had called in sick that day. This allows two interpretations of the man's telling his boss the next day at work, "That day of bedrest cured me." In one, the man knows his boss saw him, so he is joking ironically. In the other interpretation the man does not know his boss saw him, so he continues to lie. In addition, activation in the right orbitofrontal cortex increased when subjects were asked to identify words that referred to mental events, such as "think," "dream," or "plan" (Baron-Cohen et al. 1994).

Autism as a Disorder of the Ventral Processing Streams

One way to indicate that some of the ventral streams running through the temporal lobes into the orbitofrontal cortex have important responsibilities in mind reading is to show that people with severe mind-reading deficits have problems primarily in these areas. Baron-Cohen et al. (1994) gave the following list of characteristics that they stated are shared by orbitofrontal patients and autistic people: impaired social judgment, utilization behavior (they grab and manipulate any nearby object), pragmatic-discourse breakdown (misunderstanding hints, sarcasm), diminished aggression, increased indifference, decreased appreciation for dangerous situations, hyperolfactory exploration, diminished response to pain, and excessive activity. Several independent groups observed hypometabolism in the temporal lobes of autistic children (e.g., Zilbovicius et al. 2000). High-resolution PET showed decreased blood flow in the right lateral temporal and right, left, and midfrontal lobes in a group of four young adults (mean age 27.7) who were diagnosed as having had infantile autism (George et al. 1992).

All of this suggests that autism primarily involves malfunctions in the ventral cortical moiety. In many cases, this may cause the autistic child to attempt to switch the functions of the ventral moiety to the dorsal moiety (Hirstein et al. 2001). Autism is characterized by underuse of the ventral streams. In other cases, these streams may be used for abnormal purposes; for example, autistic children obsessed with certain artifacts, such as those who can identify every type of train locomotive, may be using streams that normally function in the recognition of people.

The fact that autistic people fail to look at other people or look at them in different ways suggests something is different or wrong in the ventral visual streams that perform these functions. These people often refuse to fixate on certain stimuli, especially the eyes of others, known as gaze aversion. Higher-functioning autistic subjects tended to have overly large skin-conductance responses to locking eye gaze with another person (Hirstein et al. 2001). This suggests that higher-functioning autistic people avoid locking gaze with others because they find this autonomic response unpleasant in the same way that the experience of anxiety is unpleasant. Or, looking at the eyes of another may initiate seizures, or at least the auras that come before seizures, which can sometimes allow the person to take action to prevent the seizure from progressing, such as averting gaze.

As we tested autistic children, we had them play with their favorite toys in order to measure their autonomic reactions. Two subjects brought in small plastic human figures, toy soldiers, action figures, and the like; however, the eyes had been scratched off of all the figures. Their parents affirmed that the children had a tendency to do this. This is reminiscent of the case in which the flashing eyes of a cartoon character caused seizures in children. It is frequently observed in the clinic that autistic children have a tendency to use peripheral vision when attending to objects of interest. The ventral visual streams are fed primarily by focal vision, whereas dorsal visual areas are fed by the periphery of the visual field (see section 2.2). Some of the savant skills of autistic people (such as the ability to remember all the details of a building seen only briefly) are also suggestive of a higher degree of attention in the periphery. Consistent with this, O'Riordan et al. (2001) found that autistic people are better than normal subjects at a visual search, such as when hidden figures must be found in a drawing.

How Does Emotional Mind Reading Work?

What is the connection between emotion and mind reading? When we understand the emotions of another, how do our own emotions operate to create a simulation of that person's feelings? One role of the orbitofrontal cortex in mind reading may be to produce either social or simulated emotions. Social emotions are my emotions, my guilt or embarrassment, while the simulated emotions represent the emotions of the person I am interacting with. What is the difference between an emotion that is mine and one that is simulatory? Compare this question with the question of the difference between activity in the ventrolateral motor cortex (ventral area 6) that represents my grasping movement, and activity that represents a grasping movement by someone I am looking at (Rizzolatti et al. 1996). One can find the difference between the two events by examining their causal pedigree and, one suspects, also their causal offspring. For instance, in the case of mirror neurons, the event in the motor cortex that is simulatory traces back to (ventral) visual areas, whereas the one that is not simulatory will either not trace to visual areas at all or will trace to different (dorsal) visual areas.

An emotion that is connected to a representation of a certain person's face can be made into a simulated emotion of that person. It need only be regarded as such (and perhaps actively not regarded as one's own emotion, i.e., prevented from having the full effects our own emotions have). In the case of simulated actions, the simulatory events must also be prevented from having the sorts of connections to the action-producing processes that my own brain events have, so that they do not result in actions on my part. There seem to be pathological cases, however, in which there is a failure to keep the simulated representations from causing behavior, such as echolalia, in which patients involuntarily repeat whatever is said to them. The brain also has trouble keeping simulated emotions from affecting us, which is no doubt part of the reason we find it unpleasant to be with nervous or fearful people. However, this emotional infectiousness also works to synchronize our emotions with those with whom we interact, something that has social and interpersonal functions.

5.3   Misidentification Syndromes

Our ability to recognize literally thousands of different individuals by slight facial differences is one of our brain's most amazing achievements. We normally recognize other people by their faces, but we can do it by their voices, their overall body shapes, even by the ways they walk. Sometimes we just need to identify people temporarily, such as a waitress in a restaurant whom we want to ask for our bill. We have to recognize her so we don't try to summon the wrong person, but she will soon be gone from our minds. Other people live in our mind for its entire life, such as parents and older siblings. Still others dwell there for shorter but sometimes deeply significant periods: spouses and lovers, younger brothers and sisters, close friends.

Patients with Capgras' syndrome claim that people close to them, typically a parent, spouse, or child, have been replaced by imposters. Their attitude toward the alleged impostor can vary. Most patients are suspicious at the very least. Many are paranoid and attribute ill will to the impostor. In some cases, however, the patient's attitude is positive: thirty percent of patients thought positively of the substitute (Wallis 1986). There should be no doubt about the degree of conviction; reports describe patients attacking (Thompson and Swan 1993) and even killing the "impostors" (DePauw and Szulecka 1988). One patient made three cups of tea—one for herself, one for her husband, and one for his impostor (Frazer and Roberts 1994). "If she and her husband went out she would surreptitiously return indoors to leave a note [for her "real" husband] stating where she had gone" (1994, 557).

When asked to explain what is different about the impostor, patients with Capgras' syndrome often point to some minor detail of the "impostor's" appearance (Todd et al. 1981). One patient claimed that she could tell her husband had been replaced because the new person tied his shoelaces differently, and another said that the impostor of her son "had different-colored eyes, was not as big and brawny, and that her real son would not kiss her" (Frazer and Roberts 1994, 557). Another woman "believed that she had the ability to distinguish between these various impostors by minor physical differences, such as height, skin texture and length of nose" (O'Reilly and Malhotra 1987, 264). Our patient D.S. claimed that photos of himself were not of him, but of "the other D.S." When we pointed out the resemblance this person had to him, he countered that the person in the photo had a moustache—he had not worn one since his accident (Hirstein and Ramachandran 1997). These claims have the ring of confabulation. Young is also dubious about them: "Although people who experience the Capgras delusion claim to be able to tell the real person from the impostor, they are unable to explain convincingly how they do this" (Young 2000, 52).

The patients also tend to confabulate when asked why they maintained relationships with the "impostors" at all. Our patient D.S. suggested that his father had sent an impostor of himself to take care of him. Another patient offered a similar explanation: "The patient denied VZ—his own wife—being his wife and stated that she was simply a person who had started taking care of him after his accident" Mattioli et al. (1999, 416). These authors recorded some other interesting confabulations by this patient. When they asked him what he did yesterday, he replied, "'Yesterday I went to the office and then started a project of a new machine. In the afternoon I took a swim in the lake.' It should be noted that it was January and AZ had not resumed work after the accident. Confronted with these objections he replied, 'I still work, although I am sometimes a little tired,' and 'but this is an especially mild January'" (1999, 418). The confabulations of AZ also changed over time: "When asked whether this woman (pointing to his wife) is his wife on different occasions, AZ answered: 'She is an old friend, probably I will marry her.' A year later, his answer was 'No, we are not married. We don't live together, she lives in a village close to mine.' The next year he replied, 'No, I have been married with another woman, this is a friend of mine'" (Mattioli et al. 1999, 418).

Capgras' syndrome is one of a fascinating collection of misidentification syndromes in which patients misidentify people, including themselves.

The syndromes have a wide variety of precipitating causes, including Alzheimer's disease, schizophrenia, head trauma, and cerebrovascular disease, although evidence is accumulating for a pattern of orbitofrontal and temporal lobe damage. The primary misidentification syndromes are discussed in the following sections.

Capgras' Syndrome

The Capgras' patient claims that people he knows well have been replaced by impostors. The delusion can also be directed at the patient himself; for example, a patient may even claim that his mirror image is not him. Capgras' syndrome is rare, but not unheard of; fully ten percent of one group of institutionalized Alzheimer's patients had the disorder at some point (Harwood et al. 1999). The duplication of people seen in Capgras' syndrome seems to be a subcase of a more general syndrome called reduplicative paramnesia. In 1903 pioneering neurologist Arnold Pick reported a patient who claimed there were two similar clinics, an old one and a new one, both of which were headed by Dr. Pick. Typically, the duplicated object is one's house or the hospital one is currently in.

Fregoli's Syndrome

This patient says that several unfamiliar people, despite looking different, are actually the same familiar person, such as the woman who recognized a friend of hers in every person she saw (Paillere-Martinot et al. 1994). Courbon and Fail (1927) named this Fregoli's syndrome after the Italian actor and impersonator Leopoldo Fregoli, who was famous for his ability to imitate people and was then popular in French music halls. De Pauw et al. described a patient who complained that her cousin and his female friend were stalking her: "They keep changing their clothes and their hairstyles, but I know it's them," she said. "He can look like an old man.... It's like an actor and actress preparing for different scenes" (1987, 434). This woman had suffered left frontal damage in a fall, and a CT scan showed an infarct in the right temporoparietal area. She had earlier been diagnosed as having temporal lobe arteritis in the right hemisphere. She scored poorly on tests requiring the matching of photographs of the same people taken at different times, indicating that she may also have had a visual problem. We sometimes seem to experience a mild form of Fregoli's syndrome; when we are looking for a particular person, we often tend to mistake similar-looking people for that person (Ellis and Young 1990).

Cotard's Syndrome

In Cotard's syndrome the patient says that he and/or others are "dead," or "empty." A patient of Wright et al. (1993) reported "feeling nothing inside." As Jules Cotard put it, "Nothing exists any longer, not even themselves" (in Förstl and Beats 1992, 417). In 1788, Charles Bonnet described a patient who, after what appears to have been a stroke, "demanded that the women should dress her in a shroud and place her in a coffin since she was in fact already dead" (Förstl and Beats 1992, 417). Wright et al. described a fascinating patient who alternated between Cotard's and Capgras' syndromes. He had some indications of orbitofrontal trouble: "he gave a history of olfactory hallucinations" (1993, 346).

The Illusion of Subjective Doubles
This patient claims that he has a double or doppelganger. This is different from having Capgras' syndrome for oneself, since the double is not regarded as an impostor, but as a separate, similar-looking person.

Intermetamorphosis
In this syndrome the patient says that people change into other people, sometimes right before the patient's eyes. Courbon and Tusques's (1932) original patient was a woman who said that her husband's face could change to look exactly like that of a neighbor, and that many other people took on the appearance of her son. Intermetamorphosis is rare, and seems clearly to also involve perceptual problems.

DeClerambault's Syndrome
This patient believes that a certain person, often someone famous, is secretly in love with him or her. Three groups of researchers (Sims and White 1973; Signer and Isbister 1987; O'Dwyer 1990) describe patients who exhibited both Capgras' and DeClerambault's syndromes. One woman believed that her doctor was secretly in love with her. "She knew of his feelings, she claimed, because his wedding ring was twisted around on his finger" (O'Dwyer 1990, 576). She also believed that her doctor had been replaced by an exact double, to prevent the two of them from meeting. "She reported no difference between the two except that the impostor did not smile as often" (O'Dwyer 1990, 576).

Two Components of Misidentification
As the case of intermetamorphosis indicates, some of these disorders seem to involve perceptual problems, although this must be investigated further. In Fregoli's syndrome, for instance, do the people actually visibly resemble the emotionally salient person, or is something else at work? While many patients with Capgras' syndrome do have face-processing problems, no characteristic profile of these problems has yet emerged (Edelstyn and Oyebode 1999).

    As with the other confabulation syndromes we have seen, many authors believe that there are two components to a delusion of misidentification: one approach is typical. Ellis et al. stated "first [there is] a disturbed perceptual/cognitive stage which may take a variety of forms; and second a corruption to the subsequent decision stage that is comparatively uniform across delusions" (1994, 119). Others (Alexander et al. 1979) proposed that Capgras' and reduplicative paramnesia occur when frontal lobe dysfunction is superimposed on a right hemisphere disturbance. Contrary to this, according to some, the problem is only in the first stage, and what follows is a logical response to an illogical perception. This idea seems to begin with Charles Bonnet in 1788 [although there is a hint of it in the epigraph for chapter 1 by Lettsom (1787): the patients reason tolerably clearly upon false premises]. Bonnet said; "They usually draw apparently logical conclusions unfortunately on a completely unsubstantiated premise" (Förstl and Beats 1992, 417). The logical conclusion, however, is that the person looks or seems different, not that he is an impostor or dead.

Two Routes to Perception of Faces

Prosopagnosia is inability to recognize faces. It even includes one's own face, as this testimonial shows: "At the club I saw someone strange staring at me, and asked the steward who it was. You'll laugh at me: I had been looking at myself in the mirror" (Pallis 1955, 219). It is interesting that despite being unable to consciously recognize people, some prosopagnosics register the normal larger skin-conductance response to the sight of a familiar face (Bauer 1984, 1986). Young (1998) summarized twenty studies showing covert responses of several types other than SCR to familiar faces in prosopagnosics, such as eye movements and evoked potentials. The fact that only some prosopagnosics show such responses may indicate that two different forms of prosopagnosia exist. In the first form, the damage is primarily to perceptual areas, "whereby they see faces as blobs, caricatures, all alike, etc." (Ellis and Young 1990, 240). The second type involves greater damage to memory regions, "whereby faces look normal enough but evoke no sense of familiarity" (1990, 240). Sergent and Signoret (1992) claimed that only the mnemonic prosopagnosics show covert recognition. This makes sense, since mnemonic patients are more likely to have a lesion farther "downstream," from the lesions of the perceptual patients, a lesion possibly occurring past the route leading from perceptual recognition to some sort of covert response, so that the autonomic response is intact.

Bauer (1984) and Ellis and Young (1990) have suggested that Capgras' syndrome is the mirror image of prosopagnosia, in that the Capgras' patient recognizes the face of, for example, his father (albeit as an imposter), but fails to experience the normal emotional reaction to him. "Patients with an intact ventral route and a damaged, disconnected, or temporarily impaired dorsal route could find themselves receiving all the required physical data to match stored representations of faces but at the same time fail ... to receive confirmatory evidence of their significance. As a result they may confabulate the existence of doubles, impostors, robots, etc." (Ellis 1994, 183). We can perhaps imagine the situation from the point of view of the patient: "That man looks just like my father, but I feel nothing for him, so he cannot be my father. He must be an impostor." Hence it was also suggested that these patients would lack the normal SCR to familiar faces, and several years later this was indeed verified (Ellis et al. 1997; Hirstein and Ramachandran 1997).

There is disagreement as to where these two routes are, however. As the quotation from Ellis indicates, Bauer's original approach, based on the work of Bear (1983), was to make a dorsal-ventral distinction. The mechanics of face recognition are performed in a ventral processing stream in the temporal lobes, while the warm glow of familiarity comes through a dorsal processing stream running up toward the parietal lobes. The ventral stream, according to Bauer, runs from the visual cortex to the temporal lobes. The dorsal stream connects the visual cortex and the limbic system by way of the inferior parietal lobe, Bear argued. The ventral stream was claimed to be responsible for conscious recognition of the familiar face, whereas the dorsal stream is responsible for the emotional-autonomic response to the sight of the face. All parties in the dispute agree with Bauer's first claim, that face recognition is performed by a ventral visual system, the disagreement is about where the other route is.

In our 1997 article, Ramachandran and I theorized that the double dissociation involves two ventral streams, rather than a ventral stream and a dorsal one. Our main reason for this is that it is now orthodoxy that facial processing is done in the ventral stream, while the dorsal stream specializes in peripheral rather than focal vision. Its function is to allow us to represent our nearby personal space and to navigate around both familiar and unfamiliar areas. Of interest is the report of a patient with reduplicative paramnesia with evidence of stroke damage to "the white matter adjacent to Brodmann areas 19 and 37 [that] most likely disrupted the occipitotemporal projection system, the fiber pathway of the ventral visual system" (Budson et al. 2000, 449). This patient had signs of frontal lobe dysfunction in that he performed poorly on tests of response inhibition. Encephalographic data suggest two routes within the temporal lobes by which visual information is transmitted to the amygdala, a medial route and a lateral route (Halgren 1992). The existence of the medial route is supported "by the fact that prosopagnosia is generally associated with bilateral lesions in the medial occipitotemporal cortex" (figure 5.3). "Apparently, the lateral route is sufficient to support the preserved ability to cognitively process faces ... as well as preserved skin conductance response to familiar faces ... in prosopagnosics" (Halgren 1992, 211).

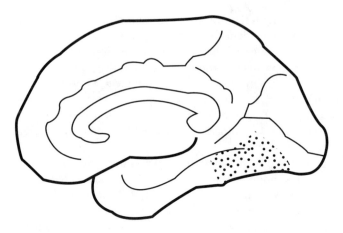

Figure 5.3
Typical site of damage in prosopagnosia. (Based on figure 5.2 in Damasio 1999. *The Feeling of What Happens.* New York: Harcourt Brace and Company.)

Recognizing faces and interpreting expressions can be doubly dissociated (Ellis and Young 1990). Working primarily from imaging studies, Haxby et al. (2000) found that the temporal lobe contains two face-processing streams and suggested that the proper way to characterize the functional difference between them is that one of them specializes in representing the changeable aspects of faces, such as direction of eye gaze and expression, whereas the other specializes in recognizing the "invariant aspects that underlie identity" (Hoffman and Haxby 2000, 80). Some writers maintained that this latter stream is specialized for faces, and so called it the fusiform face area. The first stream involves the cortex around the superior temporal sulcus; the second one is roughly on the other side of the temporal lobe, underneath and toward the middle of it, in the fusiform gyrus. Both regions receive their input from a prior processing area located in the inferior occipital gyrus (Haxby et al. 2000).

An area that represents the changing face is well situated to be part of systems for understanding eye movements, e.g., to objects of interest, and also facial expressions of emotion. "Eye gaze, in particular, is a powerful social signal that can guide our attentions and can inform us about the intentions and interest of another person" (Hoffman and Haxby 2000, 83). As early as 1982, Fried et al. showed that epilepsy patients had difficulty labeling emotions in faces if the right superior temporal sulcus was electrically stimulated. Researchers who measure evoked potentials detected by electrodes implanted in the brains of epilepsy patients before surgery found activity in the posterior superior temporal sulcus while the subjects viewed faces (McCarthy et al. 1999). When brain activity was monitored by fMRI while the subjects attended to the emotion a face was expressing, enhanced activity was detected in the right superior temporal sulcus (Narumoto et al. 2001).

Once we conceptualize the two streams as functioning in this way, dissociations of the two can be found in animals and humans. In monkeys, cells have been found in the superior temporal sulcus that respond to gaze angle and to angle of profile. Lesions in the monkey superior temporal sulcus produce impairments in perceiving the direction of gaze, while preserving the ability to judge face identity (Hoffman and Haxby 2000). Hasselmo et al. (1989) recorded activity in cells in the superior temporal sulcus and the convexity of the inferior temporal gyrus. Cells in the superior temporal sulcus responded differently to different expressions no matter which individual the monkey was viewing, and cells in the inferior temporal gyrus responded differently to different individuals no matter which expression the person showed. When the superior temporal sulcus area was ablated, the monkeys were also poor at identifying gaze direction (Perrett et al. 1992). However, the monkeys were not poor at determining identity; that is, they were not prosopagnosic. Campbell et al. (1990) tested a patient who was unimpaired in discerning the direction of eye gaze, but impaired in face recognition.

Both of these ventral face-processing routes may contact the amygdala. "The visual amygdaloid network is extended by corticocortical connections of area 12l with the entire extent of area TE in the inferior temporal cortex" (Carmichael and Price 1995b, 636). The amygdala itself responded more to emotional faces, either happy or fearful, than to neutral faces (Breiter et al. 1996). Other measurement techniques allow exact information to be obtained about the timing of processing events in these ventral streams. Streit et al. (1999) had subjects judge emotion from facial expression as they monitored activity in their temporal lobes with magnetoencephalography (MEG). They found activity in the posterior superior temporal sulcus 140–170 msec (0.140–0.170 seconds) after stimulus onset. The response was slightly later in the amygdala, at 220 msec. Given the amygdala's autonomic role (see chapters 3 and 4), these results indicate that both routes may be able to produce SCRs at that point in their processing stream. The fusiform gyrus area may be involved in producing an SCR to the sight of a familiar face, whereas the superior temporal sulcus may be involved in producing autonomic responses to emotional facial expressions. The autonomic response to familiar people might come from at least two places, the amygdala or the orbitofrontal cortex, but recall from chapter 4 that the connections and functions of these two structures are roughly similar.

In Bauer's approach, the dorsal stream presumably produces an emotional response to face identity, whereas Haxby et al. located these functions in a ventral stream. But Bauer traced a processing route moving to the superior temporal sulcus, then to the inferior parietal lobule, whereas Haxby's group showed processing running primarily in the other direction. If Haxby's two routes are the two processes that are doubly dissociated in prosopagnosia and Capgras' syndrome, this indicates that prosopagnosia involves a damaged fusiform area and an intact superior temporal area, whereas Capgras' involves the opposite. Consistent with this, our patient with Capgras' syndrome was defective in processing gaze in that he was unable to give correct answers as to whether or not a face in a photograph was looking toward him.

Autism appears to be characterized by malfunctioning or nonfunctioning of both face-processing streams. Normal subjects looking at face stimuli showed increased blood flow in both the fusiform gyrus and the superior temporal sulcus, but autistic subjects failed to show significant activity in either area (Pierce et al. 2001).

Not all sides agree on the idea of two independent streams, however. DeHaan et al. (1992) proposed a single route for face processing, but that route can become disconnected either from areas that produce autonomic responses, or from areas that produce conscious awareness of the recognized face. A modified proposal in this regard suggests that an initial pathway is involved in recognizing the face, but that two pathways diverge from there (Breen et al. 2000a). The first one runs to a system that contains semantic and biographical information about the person recognized, and the second generates the autonomic responses to familiar people. The difficulty and confusion delineating the two routes may be due to two factors. First, the routes run close by each other (at least in a ventral-ventral theory), so that many patients may have damage to both. Second, anatomical studies show that the routes interact at several points (see Saleem et al. 2000). Nature is not shy about using easy shortcuts; wherever some useful product can be gained from combining two processing streams, one suspects it will tend to happen.

5.4   A Mind-Reading Theory of Misidentification

In 1992 cognitive neuropsychologist Chris Frith suggested that a wide range of delusional disorders involve mind-reading deficits. The list included delusions of persecution and misidentification, and third-person auditory hallucinations. Since then, Frith and his colleagues have extensively tested schizophrenic patients on a variety of different mind-reading tests, finding that schizophrenics with behavioral symptoms have impairments in that realm (Pickup and Frith 2001). In this section I examine the feasibility of explaining misidentification syndromes as due to a malfunction of different mind-reading capacities. This is part of my larger goal of showing that confabulation syndromes include mind-reading problems.

Think of how much effort your brain devotes to representing the people you know well. You can recognize their faces at all angles, in poor light, and at some distance. You can recognize their voices no matter what the emotional tonality. You can recognize their walks, their laughs, even their characteristic scents. These are all what we might call external representations of those people. You also represent their minds—you have internal representations of them. Looking at the face of someone you know well is a quite different experience from looking at the face of a stranger. The difference has to do with the way that you attribute a mind to the person you know. You attribute characteristic emotions and moods, even beliefs. Imagine the following scenario: Your friends have set you up on a blind date; well, not exactly blind because they have a picture of your date. Unfortunately, it is a rather bland shot of her smiling at the camera. As you look at the picture, you try to form hypotheses about her personality, but you find that you just can't get a fix on her. You can't tell whether she is outgoing or introverted, patient or impatient, open-minded or dogmatic, intelligent or stupid. Once you meet her, though, and can see how she animates her face, additional processes will be going on in your brain when you look at her, and even when you look again at that same picture. Ellis et al. used the term *personality* to describe this internal component: "In the intermetamorphosis delusion there is a simultaneous alteration in both the appearance and personality of someone else. In the Fregoli delusion it is just the personality that changes" (1994, 118).

Impostors are people pretending to be someone else. They normally perpetrate their deception by their strong visual resemblance to the person they are pretending to be. They look the same as the people they are mimicking on the outside, but are completely different on the inside. Some comments of patients with Capgras' syndrome refer straightforwardly to mental features. One such patient "became convinced that her husband's personality had changed" (Frazer and Roberts 1994). Another claimed there were two doctors looking after him, each with a different personality: "The first consultant (who he called John Smith) was 'a nice bloke,' whereas the second (a Dr. J. Smith) was someone who was 'distant and aloof'" (Young 1998, 39). Some reports of patients who fail to recognize their own mirror reflections seem to show changes in the internal representation of the patient himself at work. "He developed the delusion that his facial reflection in mirrors and other glass surfaces was no longer his own face but the work of an autonomous evil being. The image of his face was identical to his own face except that the image appeared to be very sinister and aggressive" (Silva et al. 1992, 574–575).

I suggested earlier that the claims by Capgras' patients that they can detect the imposter by slight differences in appearance might be a kind of confabulation. Breen et al. (2000b) present a nice description in which author Clifford Beers recounts his experiences after his recovery from Capgras' delusion. When relatives came to see him, he said, "I was able to detect some slight difference in look or gesture or intonation of voice, and this was enough to confirm my belief that they were impersonators" (Beers 1953, 62–63). Beers thought he detected slight differences, as many patients do, but even during the delusional phase, he regarded these as confirmatory of a preexisting belief that the person he was looking at was a different person, some stranger who was "impersonating" his relative.

The relations between the internal and external representations are complex and can malfunction in several different ways. When we simulate another person, we contain his or her mind within our own. A woman with DeClerambault's syndrome believed that a famous conductor was in love with her. "She believes she hears the conductor's thoughts in her brain, although these thoughts are not experienced as auditory hallucinations (she cannot identify a voice as such), and the conductor also knows her thoughts as if by telepathy. Some difficulty arises because they 'share a brain,' making it particularly difficult for one or the other of them to concentrate when the other is reading" (Kopelman et al. 1995, 73). The young woman with Fregoli's syndrome who recognized a friend of hers in every person she saw also "felt that this particular male friend was inside every person. She also felt that her dead grandfather was living inside a nurse" (Paillere-Martinot et al. 1994, 201). This containment metaphor was also used in an explanation of Fregoli's syndrome: "Here the delusion lies in the belief that people who themselves are usually unfamiliar to the patient, may, in reality, contain within someone with who s/he is familiar" (Ellis et al. 1994, 118).

Areas of the mind-reading system such as the orbitofrontal cortex and the temporal lobes have been found to be damaged in patients with Capgras' syndrome. One of the most frequent phrases in case reports of patients with Capgras' syndrome is "bifrontal damage" (e.g., Benson et al. 1976; Joseph et al. 1990). Two-thirds of a group of twenty-three patients showed clear cortical atrophy on their CT scans, and in all of them the signs were bilateral; in some the damage was frontal, and in others it was in parietal or temporal areas (Joseph et al. 1990). Many patients with Capgras' syndrome have suffered closed head injuries. The orbitofrontal regions are frequently damaged in closed head injuries because of the way they lie against the inferior surface of the skull (Starkstein and Robinson 1997). "Damage can arise from the collision of the brain with the solid internal surface of the skull, especially along the jagged surface of the orbital region" (Gazzaniga et al. 2002, 121). When Holbourn (1943) filled a skull with gelatin and violently shook it, the orbitofrontal area alone was torn and "chewed up." Two of three patients of Benson et al. (1976) with reduplicative paramnesia were car accident victims, as was ours with Capgras' syndrome (Hirstein and Ramachandran 1997). Head-on collisions in which the head hits the steering wheel or dashboard may cause orbitofrontal damage.

With the increasing availability of MRI, more specific reports of damage areas are beginning to emerge, including studies that report orbitofrontal damage. In one patient, MRI showed lesions in the right hemisphere in the middle and inferior frontal gyri, orbital gyrus, gyrus rectus, and the superior and middle temporal gyri (Mattioli et al. 1999). In the left hemisphere, the damaged areas were the gyrus rectus and paraterminal gyrus, together with the adjacent portion of the anterior commissure (figures 3.2 and 4.1). This patient claimed that his wife was an impostor, that his house was not in fact his house, and confabulated frequently in personal recollections and on a test of long-term verbal memory. As with other confabulators, "when confronted with reality, he would reluctantly admit it for a moment without any sign of surprise or insight, but soon would go back to his former interpretation" (Mattioli et al. 1999, 417).

There are also frequent reports of temporal lobe damage in Capgras' syndrome. In a review of the neuropsychological findings in the syndrome, neuroimaging results indicated "a link between [Capgras'] and right hemisphere abnormalities, particularly in the frontal and temporal regions" (Edelstyn and Oyebode 1999, 48). An epileptic patient with a seizure focus in the right temporal lobe developed Capgras' syndrome after his seizures (Drake 1987), and three patients with intermetamorphosis had temporal lobe epilepsy (Bick 1986).

Metabolic uptake in different brain areas was measured in patients with Alzheimer's disease with and without a delusional misidentification syndrome (Mentis et al. 1995). Patients who made misidentifications showed hypometabolism in the orbitofrontal (Brodmann's areas 10 and 11) and cingulate cortices (areas 32 and 25) bilaterally, and in the left medial temporal cortex. Curiously, they also showed hypermetabolism in superior temporal and inferior parietal areas.

Several writers suggested that right hemisphere damage is more frequent than left hemisphere damage in Capgras' and the other delusional misidentification syndromes (Fleminger and Burns 1993; Förstl et al. 1991; Hakim et al. 1988). One patient with a vascular lesion in the right frontal region had a memory impairment, left-side inattention, and reduplicative paramnesia (Kapur et al. 1988). Another had severe right hemisphere damage and reduplicated people, including himself, buildings, and his cat (Staton et al. 1982). Alexander et al. (1979) seemed to put their fingers on the problem areas when they claimed that Capgras' syndrome is produced by a combination of right temporal and bilateral frontal damage. Signer

(1994), who reviewed 252 cases of Capgras', posited a combination of left temporal and right frontal damage.

Many writers have argued that in Capgras' syndrome there is a disconnection between the representation of a face and "the evocation of affective memories" (Ellis and Young 1990, 243; Christodoulou 1977; Alexander et al. 1979; Staton et al. 1982; Bauer 1986; Lewis 1987; Stuss 1991; Hirstein and Ramachandran 1997). Patients recognize a face, but do not feel the expected emotions and confabulate to explain this: "When patients find themselves in such a conflict (that is, receiving some information which indicates the face in front of them belongs to X, but not receiving confirmation of this), they may adopt some sort of rationalization strategy in which the individual before them is deemed to be an impostor, a dummy, a robot, or whatever extant technology may suggest" (Ellis and Young 1990, 244). Mentis et al. (1995) pointed to problems both in sensory areas and in the orbitofrontal cortex in the misidentification syndromes, but spoke only generally about "sensory-affective dissonance". "The patient perceives the stimulus, but not its emotional significance and relevance to the self" (1995, 447). Constructing a "solution" to explain their strange experience seems to help these patients regain their cognitive equilibrium. The misidentification patient "rationalizes" a conflict between present "feelings of unfamiliarity and any preserved memories for a person by splitting the person's identity with an invented double" (Mendez 1992, 416). Capgras himself noted that with the onset of the delusion, feelings of strangeness and unreality might disappear (Capgras and Reboul-Lachaux 1923).

A mind-reading deficit would include these sorts of emotional deficits, but this approach emphasizes the failure of the patient's brain to produce the right simulatory emotions. It may be that both the patient's own emotional reactions and the emotions he attributes to others have changed. If it is true that we use our own emotional system in simulation mode to understand the emotions of others, a simultaneous change in both these realms is less surprising. One argument for the theory of mind approach to the misidentification syndromes is that the disorder often has broad effects on the patient's theory of mind. Many misidentification patients attribute their own paranoid state of mind to those about whom they have the delusion. One patient, who alternated between Capgras' and Cotard's delusions, accused the nursing staff of having murdered members of his family (Butler 2000). This man was "fearful, and perseverated on themes related to death" (2000, 685). When the patient interacted with his father, he "minutely examined [his father's] face before accusing him of being a 'criminal' double who had taken his father's place" (Butler 2000, 685). Wright et al. (1993) described a patient who exhibited Cotard's syn-

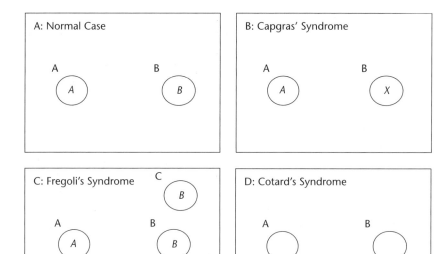

Figure 5.4
Internal and external representations in the misidentification syndromes. (*A*)
The normal case, seen from the point of view of A. (*B*) Capgras' syndrome; the
subject believes that the target person has a familiar external appearance, but an
unfamiliar mind. (*C*) Fregoli's syndrome; the subject sees several people who
have the mind of someone he knows well. (*D*) Cotard's syndrome; the subject
believes that he and/or others are "dead" or "empty."

drome when depressed, but exhibited Capgras' syndrome when he experienced persecutory delusions.

Here is how the mind-reading theory explains the different misidentification syndromes. In Cotard's syndrome, the internal representation system is not working (figure 5.4). This is why the experience for the
patient is that of looking at an empty shell. In Capgras' syndrome, an internal representation is at work when the patient looks at the "impostor,"
but it is not the same one that the patient has always used. The normal internal representation has either been damaged in some way, or a different
one is activated by the sight of the person. Fregoli's syndrome occurs when
the same internal representation is activated by the sight of several different people, or in extreme cases, by the sight of anyone.

A relationship may exist between these phenomena and the distinction we saw between the medial and orbital networks in the orbitofrontal

cortex (Carmichael and Price 1996). Recall that the medial network had extensive autonomic connections, including connections to the anterior cingulate, amygdala, and hypothalamus, and was postulated to be involved in attaching an affective value to representations. The orbital network, in contrast, lacked these limbic connections and instead had connections to areas specializing in the different sensory modalities, as well as extensive connections with the dorsolateral frontal cortex. Since the inferior and medial temporal lobe streams specializing in face identity connect more to the orbital network (area 12l, figure 4.4), that network may specialize in representing the seen person's face. The lateral temporal lobe streams connect to both a lateral orbital area (12o) and a medial area (13a). The lateral area may specialize in representing the facial expression of emotion and the medial area may produce a simulation of the emotion registered by that face. When it is being used for simulatory purposes, the medial network functions as what I have been calling an internal representation, while the orbital network functions as an external representation, in this hypothesis.

Let us look briefly at other theories of Capgras' syndrome and discuss considerations for and against them.

The Underidentification Theory

Christodoulou (1977) and De Pauw et al. (1987) contended that Fregoli's syndrome involves *hyperidentification*, in that people the patient knows tend to proliferate—strangers are seen as familiar. Capgras' syndrome, in contrast, is a species of hypoidentification, because people the patient knows disappear—familiar people are seen as strangers. Feinberg and Roane (1997) similarly held that it is useful to think of some misidentification syndromes as involving underidentification; failure to acknowledge what is familiar, as in Capgras', and overidentification, treating a stranger as a familiar person, as in Fregoli's or DeClerambault's syndrome. This approach is different from the theory of mind approach, but not inconsistent with it. The patient with Capgras' assigns an unfamiliar mind to "impostors," and those with Fregoli's and DeClerambault's syndromes assign a familiar mind to strangers. The familiar-unfamiliar dimension cannot accommodate the other misidentification syndromes, however, such as Cotard's. Another disconfirming piece of evidence for this hypothesis is the patient with both Cagpras' (a sign of underidentification) and intermetamorphosis together with false recognition of unfamiliar faces (signs of overidentification; Edelstyn et al. 2001).

The Mnemonic Theory

Some patients with Capgras' syndrome have odd memory problems. Staton et al. (1982) proposed that Capgras' is fundamentally a memory disorder in which the patient fails to relate current experience of a person to earlier episodic memories of that person. They compared this problem with memory problems in Korsakoff's, which they suggested is a more general version of the same dysfunction (see also Pisani et al. 2000). Hudson and Grace (2000) provide a specific version of this memory theory, suggesting that misidentification syndromes are caused by disconnection of face (identity) perception areas from long-term memory systems in the temporal lobes that contain processes responsible for retrieval of face information from long-term memory.

Some memory theorists may be mistaking confabulations for accurate memory reports, however. In the mind-reading theory, something has changed, but it is something that we are normally not explicitly aware of—an internal representation. The patients reason that there must be something externally different about the person, and confabulations about minor physical differences arise from this. One patient said that he believed things had changed because they looked different than they did in the past, and the authors interpreted that to mean that he has "difficulty comparing and integrating present cues with past experiences" (Staton et al. 1982, 30). The remarks he made may actually be confabulations, however, created in an attempt to understand why people, pets, and buildings seem different.

There does seem to an odd memory-individuation problem in Capgras' syndrome. When we were testing our patient for his ability to discern whether or not a model in photographs was looking at him, he misidentified the model as three different people. These memory phenomena are not inconsistent with the mind-reading approach; it may be that the memory impairment causes substitution of one internal representation for another.

The Identity Theory

According to Cutting (1991), Capgras' syndrome involves a categorization failure, specifically, a problem recognizing the identity of *unique* individuals. This idea is based on Kosslyn's (1987) claim that whereas the left hemisphere specializes in recognizing the general category a percept falls under, the right hemisphere specializes in recognizing individuals within a category. Some theories of the misidentification syndromes use the concept of identity to describe what is missing or has changed. Ellis and Young's (1990) model includes "face-recognition units," and different "person-identity nodes" that exist in semantic memory, where known biographical information about that person is stored. Surely such person concepts exist, but their semantic information is a minor part of our representations of others. We might imagine two quite different people about whom we have stored largely the same biographical information. This similarity in biographies does not cause us to regard them as similar people. We do not represent people's personalities in semantic memory by recording the fact that

Jan has a bad temper, for instance. Rather, we model Jan's mind when we interact with her by modeling her emotions with an internal representation of her mind.

The reader may have noticed the reference to the man who "reduplicated" his cat. One challenge for the theory of mind approach to misidentification disorders is the fact that they can be directed at a pet. Patients can also misidentify inanimate objects, such as houses and other possessions, a fact that seems contrary to the mind-reading approach. One also has to explain why a misidentification disorder such as Capgras' happens only with people the patient knows well. I will try to respond to these three challenges to the mind-reading hypothesis, which I will call the dog problem, the house problem, and the friends and family problem.

The Dog Problem

A twenty-three-year-old woman claimed that her cat had been replaced by the cat of a former boyfriend (Ehrt 1999). This problem is the easiest to respond to, and in the end it strongly confirms the idea that the misidentification disorders are theory of mind disorders. The main reward from owning pets is that we see that they have minds and genuine personalities. It is a truism to say that people become quite attached to their pets; the animals become members of the family. And we are all familiar with people who attribute perhaps too much of a mental life to their pets. This yields a prediction: Patients with Capgras' syndrome who include animals in their delusions have had a close association with the animals, and have attributed a mind and personality to them. In support of this, an fMRI study showed that in humans, both the fusiform gyrus and superior temporal sulcus respond the same to faces, animals, and faceless animals (Chao et al. 1999). This fact explains "why difficulty identifying four-legged animals is a common, perhaps the most common, co-occurring symptom in prosopagnosic patients" (Chao et al. 1999, 2949).

The House Problem

A woman claimed that both her husband and her house had been replaced (Frazer and Roberts 1994). A man's Capgras' syndrome was directed at his wife and he also complained that, "this house is not my house" (Mattioli et al. 1999). A lively debate has arisen about whether the brain has special equipment used solely for perceiving faces. Behrmann and Moscovitch (2001) insist that no one has ever described a prosopagnosic patient who did not have additional perceptual deficits involving objects. They also note that a "large proportion" of prosopagnosic patients are achromatopsic and have topographic agnosia, a fact that militates against any attempt to use the existence of prosopagnosia itself to argue for face-specific areas. A patient who reduplicated visual stimuli, including people and buildings, was impaired in his ability to recognize faces from memory, and performed very poorly at recognizing emotions in faces (Young et al. 1994). He was very bad at matching two photographs (from different angles, with different lighting) of the same unfamiliar person.

I suggest that we form internal representations for more than people. Imagine that you grew up in one of those early suburbs where all the houses are built according to the same blueprint. You stand in the street looking at your house and the two houses on either side of it. Externally, your house looks almost exactly like the others. But for you, looking at your house is a completely different experience from looking at the ones nextdoor. The difference is not due to features of the outside of the house, but rather to your beliefs about what is inside your house, and your emotional memories of events that took place there. When fMRI responses to houses were recorded, they activated an area of cortex adjacent to the part of the fusiform gyrus that responds to faces, animals, and faceless animals (Chao et al. 1999).

Philosopher Henri Bergson argued in the early 1900s that we have two ways of understanding, and that these two ways are applied to objects as well as people. As Kolakowski put it: "We get to know a thing, Bergson says, either by circling around it, or by entering into it. If we stay outside, the result depends on our standpoint and is expressed in symbols [Bergson called this analysis], whereas in the second kind of cognition we follow the life of the thing in an empathic identification with it [Bergson called this intuition]" (1985, 24). We even apply intuition to objects. For instance, if we see a piano fall out of a moving truck and smash on the pavement, we cannot help feeling squeamish, as if we are imagining what it might be like to experience that ourselves. If Bergson is correct, this shows that the abilities we use are not applied only to humans, but are generalized to all objects. The existence of animism among primitive peoples (in which trees, clouds, mountains, and so on are given human attributes) lends support to this idea. We naturally tend to use intuition on everything, at least until we become convinced that doing so is not effective.

The Friends and Family Problem

Ellis and Young (1990) suggest that Capgras' syndrome occurs only with faces that have a particularly strong affective component associated with them. Thus there is a certain threshold below which the patients' emotional responses are not strong enough for them to notice their absence. However, this phenomenon may indicate which particular theory of mind processes are compromised in the misidentification disorders. There are "shallow" processes we can use for anyone, but we construct deeper, more thorough simulations of people we know well. We have more thorough simulations of their beliefs, but more important than this are detailed models we form of their particular emotional reactions to different stimuli. Getting to know someone involves getting to know her likes and dislikes, what makes her happy, what makes her angry, and so on.

One of the biggest problems for this view has always been how we manage to simulate people who are very different from us. One way to deal with this is to posit the existence of less-detailed, generic internal representations of character types we employ to understand strangers quickly. During the course of our lives we become aware of a set of distinct character types; for instance, a strong-willed character, energetic and hard driving, assertive and unself-conscious. We make different predictions about what a person of this sort will do when confronted with an obstacle, for instance. Our knowledge of these types is partly simulatory; we understand what it would be like to be someone with that character. We often fit people we meet into a type to better understand them, although we are also often mistaken. As we become older and wiser, we add to this repertoire of character types so that we can apply them to more people and make more accurate predictions about what they might do. The Capgras' patient may be employing one of these generic internal representations when he looks at the "imposter," rather than the specific internal representation he normally uses.

Developmental Evidence
Brain damage allows us to see the human mind in a particularly naked form, but these moments can also occur as the mind develops. Many problems experienced by patients with brain damage are also seen in children. Most children are able to understand the false beliefs of others around age three or four (Johnson and Wellman 1980). When they are around six or seven, they become capable of understanding second-order false beliefs, that is, beliefs about the beliefs of others (Perner and Wimmer 1985). At ages nine through eleven children develop the ability to recognize faux pas (Stone et al. 1998). The performance of patients with right hemisphere damage on mind-reading tasks is similar to that of normal three-year-olds who have not developed the ability to achieve such mind-reading functions (Joseph 1998). Learning about emotions occurs later than learning language, for instance. In this process, adult guidance is especially important. Often situations merely arouse us in a nonspecific way, and, as William James suspected, we can be pushed into any of a number of emotions, depending on the context. The particular emotion we are pushed into depends especially on adult examples. "Mechanisms for recognizing emotions may be closely linked to those involved in experiencing the equivalent emotions because this has evolved as one of the ways we learn our emotional reactions; many of the things which frighten us do so not be-

cause we have direct experience, but because we have learnt from the reactions of others" (Young 1998, 25).

The inability of patients with anterior cingulectomy to distinguish fantasy from reality (see section 3.4), and the trouble sociopaths have with being aware that they are lying (section 4.4) remind one of similar behavior in children. The normal mind internalizes checking, inhibiting forces, which come originally from the parent. What stops children from telling stories is that these stories are met with increasingly powerful negative reinforcement. Parents become alarmed by a story, and when they find out that it was made up tend to make it quite clear to the child that such behavior is not acceptable. Similarly, what makes children consider the feelings of others is that selfish actions are negatively reinforced. "How would you like it if Sally did that to you?" we ask. Children learn to simulate the minds of their parents by simulating the inhibition their parents exercise on them, that is, their parents' admonitions. Some of us have internal representations of people we knew in the past, often a parent or an influential teacher, that play an inhibitory role in our thinking. As we contemplate doing something, these mentor figures speak up, sometimes from the grave, to warn us against it. Sometimes we even hear their tone of voice or turn of phrase.

5.5 Conclusion

I think the mind-reading theory of Capgras' syndrome makes it far less bizarre. The Capgras' patient is looking at someone who visually resembles his father, but who appears to have a different mental life, a different personality, with different dispositions to do different things. This is exactly what an impostor is, and this is exactly the experience one would have looking at an impostor. The mind-reading theory also explains why DeClerambault's syndrome is so often lumped with misidentification disorders, although the patient does not misidentify the target person, but systematically attributes the wrong mental states to the person. Similarly, Cotard's syndrome does not, strictly speaking, involve misidentification, but it does seem to involve a failure to attribute a mind to the person in question. Perhaps in Capgras' syndrome the patient's representation of an individual's mind is destroyed, while the face representation is intact. However, representations of other minds are available to be used. (If all the representations of minds are destroyed, Cotard's syndrome results: "Those people are empty, dead, shells.") When this other mind representation is associated with the familiar face of an individual, it creates the impression of an impostor.

Interpreted this way, misidentification disorders are analogous to memory syndromes such as Korsakoff's and ACoA aneurysm. In both cases the patient has a disability he does not know he has. The primary realm in which the subject confabulates is the realm where proper formation of beliefs relies on the damaged area. To form and maintain beliefs about what happened to me in the past (i.e., memories), certain brain processes have to be operating properly. To form beliefs about the people I am interacting with, other brain processes have to be operating properly.

What do we do with the mind-reading system when we are not with someone? One of the main checking processes we use is to try our idea out on someone else and see what he or she says. Other evaluation processes are merely internalized versions of these: We imagine how a certain person (or type of person) would react to our current thought or intention. The inhibition that prevents confabulation may be based on the response of an imagined audience to the thought or the rehearsal of the planned action. Rehearsing an action is often part of the process of forming an intention to do that action. However, the intention-forming and action-executing processes can be halted if the planned action fails certain tests. For instance, it might violate certain social norms. At dinner, I need the salt and I spy it across the table next to my friend's plate. The impulse to reach over and grab it pops up, but before it can crystallize into an intention, I consider it from my friend's point of view, and this causes me to inhibit that particular action and invoke a standard remedy—ask her to pass the salt. We interpret the emotions of others in such a way that we internalize them, where they work as inhibitions to thoughts and actions. The mind-reading system does more than read minds. It has a self-mode, and an other-mode, so to speak.

Confabulators share certain traits with sociopaths and with autistic people, if the hypotheses presented here are correct. Neither confabulators nor sociopaths show awareness that they have a problem. Neither adequately represents the minds of people with whom they interact. If the connection between confabulation and mind reading holds, autistic people who are capable of communicating should be confabulatory. They do have trouble with inappropriate social behavior, as would be predicted. They are also disinhibited. High-functioning autistic people often ask inappropriate questions, for instance. Because all of these conditions may involve orbito-frontal malfunction, there should be an interesting overlap in the symptoms of autistic people, sociopaths, and confabulators. Both autistic people and sociopaths fail to show the proper emotional responses to the distress of others. There is also a relation here to the idea that OCD is the opposite of being confabulatory, since some of these patients are often plagued by social emotions such as guilt, or are overly self-conscious. Confabulators are markedly unself-conscious and show no guilt at all. As we will see in the next chapter, they calmly deny they are paralyzed when they are, or they may even deny ownership of their own arms, while showing no concern about how these denials sound to their listeners.

6 Unawareness and Denial of Illness

You know Harpastes, my wife's fatuous companion, has remained in my home as an inherited burden.... This foolish woman has suddenly lost her sight. Incredible as it might appear, what I am going to tell you is true: She does not know she is blind. Therefore, again and again she asks her guardian to take her elsewhere. She claims that my home is dark.
—Lucius Seneca (c. 55 A.D.)

6.1 Denial

Damage in certain places in the brain can produce not only disabilities, but strangely enough, complete unawareness of those disabilities and outright denial on the patient's part that anything is wrong. Babinski (1914) coined the term *anosognosia* for lack of knowledge, or unawareness of, illness. It is formed from the roots *a* (un or without), *noso* (disease or illness), and *gnosia* (knowledge). This unawareness seems to play a role in causing denial in some patients. They calmly but consistently deny that they are paralyzed, blind, or affected in any way by brain damage. Often, when asked why they cannot move their left hand, for example, they confabulate about how they are tired, their arthritis is acting up, and so on. Denial of left-arm paralysis typically disappears a few days after a stroke. This use of "denial" is semitechnical, but is obviously related to our everyday uses when we say such things as, "His parents are in denial about his drug dealing," "He is in denial about his unfaithful wife," and so on.

Damage to parts of the right hemisphere or other areas can cause a type of hemianopia, in which vision on the left side is lost, as well as hemiplegia, great weakening or paralysis of the left arm and/or leg. Sometimes people with one or both of these conditions neglect the left side of their personal space, the nearby surrounding space that we perceive and direct actions into. They neglect it in that they do not notice, report, or respond to anything in the left visual field. Such patients seem to "forget" that that side of space exists. They do not visually explore that side, even with their remaining visual abilities, nor do they spontaneously orient to objects there. Such a patient may crash into objects on her left side, fail to eat the food on the left side of the plate, or even fail to dress or wash that side of her body.

The quick diagnostic sign of neglect is the line bisection test (Albert 1973). The patient is shown a line drawn horizontally on a sheet of paper so that half the line falls to his left and the other half to his right; he is then asked to draw a line through the center of it. His bisection will be far to the right of center, since he is bisecting only the part of the line that he is aware of, the right side. Damage to the analogous area in the left hemisphere tends to be much less disabling to our spatial functions on the right side, apparently because the right hemisphere also represents the right side of space and can accomplish the necessary representation and computation tasks on its own.

Some patients deny that the paralyzed limb is in fact their own, a condition known as asomatognosia (lack of body knowledge). Bisiach and Geminiani recorded some interesting confabulations about this: "The patient may claim that the contralesional arm belongs to a fellow patient previously transported by ambulance, or that it had been forgotten in bed by a previous patient" (1991, 20). Patients may also display hatred of the paralyzed limb, a condition known as misoplegia. It is interesting that the irrationality can expand beyond the left arm to cover the entire left side of space. The patient may claim that the arm belongs to another person who is lying at his side. One patient claimed that his daughter was lying on his left side, making sexual advances (Anton 1899).

The other type of illness that is denied is, strangely enough, blindness. The first detailed description of anosognosia was offered by von Monakow in 1885. He reported a patient who failed to realize that he was blind after cortical damage. Anton described a similar case in 1899, and denial of blindness came to bear the name Anton's syndrome. Anton's original patient could not see on his left, could not feel anything on his left side, and never moved his left limbs intentionally, after a vascular lesion in the posterior part of the right hemisphere. In 1915, Pick described a thirty-one-year-old alcoholic patient who had a vascular lesion with blindness on his left side and great weakness of his left arm, both of which he denied.

The percentage of paralyzed people who deny illness is surprisingly high. Of 100 hemiplegic patients, 28 denied illness; of the rest, 48 had full awareness of their problem, and the remaining 24 had left hemisphere lesions with aphasia severe enough to rule out the standard method of testing for denial—simply asking the patient whether he can move his arm (Nathanson et al. 1952). Twelve of thirty-six patients with paralysis had anosognosia, and the disorder was most frequent among patients with left visual field defects—twenty-eight of thirty-two (Bisiach et al. 1986). It is known that denial of hemiplegia and denial of hemianopia are dissociable, but no anatomical basis for the difference has yet been found. In addition, there are hemiplegic and hemianopic patients with large lesions in the inferior temporal cortex who do not deny their problems (Bisiach and Geminiani 1991). A fifty-seven-year-old patient with left hemiplegia and left hemianopia was described as follows: "He was disoriented for time and place, could not remember his doctors' names, and actively denied any physical disability. When asked if he could walk or dance, he would immediately say yes; when asked to raise his arms or legs, he would raise the right limbs and insist that both arms or legs had been raised" (Stuss and Benson 1986, 11). This man also had a trait we have seen in other confabulating patients, a tendency to admit the truth when confronted with it, but lapse back into his confabulations soon after: "When his hemiplegia was demonstrated to him he would accept the obvious fact and repeat the examiner's statement concerning the cause of the disability, but within minutes, if asked whether he had any disability, would adamantly deny hemiplegia" (Stuss and Benson 1986, 111).

Anosognosic patients have abundant evidence that they are paralyzed or blind. Minimally, they should know that their knowledge may be compromised in that realm, so that they should answer "I don't know" or "I'm not sure" when asked about their problem areas. Nevertheless when asked, these patients "may suddenly appear absent minded, their attention is distracted elsewhere, or they move the unaffected limb on the other side. Often they make such statements as 'Here you are,' implying the purported execution of the required movement" (Bisiach and Geminiani 1991, 19–20). A favorite confabulation appeals to the left side's basic inferiority in right-handed people: "In the exceptional instance in which the patient admits the lack or insufficiency of movement, he may sometimes explain it in various ways as being due to the fact that the involved limb is the left and therefore weaker one" (Bisiach and Geminiani 1991, 20).

These denials, and other remarks the patients make to support them, are confabulations, according to several writers (Geschwind 1965; Joseph 1986; Ramachandran 1995). Some writers use the phrase "confabulated denials of illness" (e.g., Gainotti 1975). The tendency to confabulate is also present in other domains in some denial patients. Patients with anosognosia and hemianopia are confabulatory about the identity of an object held in their left visual space (Feinberg et al. 1994).

Patients in denial have a curious lack of emotionality, even when discussing their conditions. Fulton and Bailey (1929) described the demeanor of a group of denial patients as that of "fatuous equanimity." Weinstein and Kahn, in their early authoritative work (1955), said that patients with complete explicit denial were usually bland and affable during interviews. They described the odd calmness of their own denial patients: "Anxiety, in the ordinary clinical sense, was not present,... Anosognosic patients seemed to maintain a serene faith that they were well which remained firm despite all disbelief by others" (1955, 17). The depth of denial comes in degrees. Some patients admit paralysis but seem completely unconcerned about it, a condition called anosodiaphoria (Critchley 1953). "The degree of anosognosia ranges from minimization of a specific deficit, indifference, or unconcern to lack of recognition and frank denial" McDaniel and McDaniel (1991).

One of the strangest aspects of neglect is the extent to which the left side of space is gone from the person's world. These patients even failed to describe the left side of scenes recalled from memory (Bisiach and Luzzatti 1978). Curious about the nature of the patient's attitude to the space on his left, researchers held a mirror on the patient's right (fully functional) visual field so that it showed the patient his left side of space. As the patient watched, an assistant then held a pen in the patient's left visual field, so that the pen was clearly visible in the mirror. The patient was then instructed to grab the pen. Amazingly, many neglect patients repeatedly bumped their hands into the mirror, or tried to reach around behind it, rather than reaching back toward their left to grasp the pen, a phenomenon Ramachandran et al. (1997) called mirror agnosia. When they asked one patient, a retired English professor, where the pen was, he replied, "It's through the looking glass." The patients seem to appreciate that the mirror reveals a space of some sort, but they do not understand their relationship to it.

An astonishing finding was that if one uses cold water to irrigate the left ear of some patients with neglect and denial, both symptoms temporarily disappear (Cappa et al. 1987). It is known that the cold water works to stimulate the vestibular system, but whether this causes the temporary remission and in what way is not known. The reasoning of the neglect patient seems to contain other curious loopholes. If patients who denied paralysis were given a nonthreatening reason for being paralyzed, the denial abated, as shown by an injection experiment. A doctor approaches the bedside of a patient denying left arm paralysis and says, "I'm going to give you an injection that will briefly cause your arm to be paralyzed. The paralysis will take effect right after the injection." After a fake injection, these patients admitted paralysis. The perfect comparison test for this is to conduct the same procedure with the other arm. When this was done, the patient typically tried to move her arm, succeeded, and said something such as, "I guess it didn't work" (Ramachandran and Blakeslee 1998).

What is the relationship among confabulation, anosognosia, and denial of illness? Specific acts of denial are confabulations. Anosognosia traditionally includes denial by definition (Stuss and Benson 1986), but it seems possible for someone to lack awareness that he is ill, yet not be able to express a denial of it; thus it would be more accurate to say that anosognosia is a lack of awareness of a deficit that may lead to confabulation. There is no doubt that a single lesion can produce both neglect and denial. The study of the anatomy of denial of hemiplegia and hemianopia merges with the problem of the anatomy of unilateral neglect according to Bisiach and Geminiani (1991).

There is still vagueness regarding what it means to be aware of one's deficit, however. A patient might admit paralysis, but revert back to denial seconds later. Several questions about anosognosia remain unanswered, including whether the patient is simply lying. This question is important here even more than in the other types of confabulation, because of the obvious motives, such as a desire to suppress the horrifying and depressing thought that one is partially paralyzed, as well as a need to present oneself to others as whole and sound. Another troubling fact is that what patients say about their illness and what they do can be inconsistent. Patients who verbally deny being paralyzed do not object to being confined to bed (Bisiach and Geminiani 1991). "In contrast, patients who verbally admit of paralysis on one side may attempt to stand and walk, or ask for tools they are patently unable to handle as they did prior to their illness" (Bisiach and Geminiani 1991, 19).

Representations of the Body

We are beginning to understand which parts of the brain embody awareness of our bodies. The case of phantom limbs after amputation shows that sensations of our body parts can be just as vivid when the part is not there. Also, the presence of phantom limbs in people with congenitally missing limbs seems to argue for the idea that some aspect of body image is innate (Ramachandran and Hirstein 1998). The functions of areas that represent the body in action are complex. Minimally, "one must develop a representation of the body, and this representation must be continuously modified by expectations (feedforward) and knowledge of results (feedback)" (Heilman et al. 1998, 1903). Information from the body arrives at the brain from two separate routes, the dorsal column–medial lemniscal system and the anterolateral system. The dorsal column system transmits primarily information about touch and the position of the arms, and the anterolateral system primarily transmits information about pain and sensations of heat and cold. This information passes through different brainstem nuclei and through the thalamus to its ultimate destination—a set of somatotopic maps located in the parietal lobe of the cortex, just behind the central sulcus, known collectively as the primary somatic sensory cortex.

This cortex is divided into four somatotopically organized areas— Brodmann's areas 1, 2, 3a, and 3b—in anterior parietal cortex (figure 2.1). Each area has a somewhat different specialty. Most inputs to area 1 are from receptors located in the skin known as rapidly adapting cutaneous receptors, which are used to identify stimuli by touch. Area 2 receives input primarily from deep pressure receptors, and area 3a from muscle stretch receptors that transmit information about limb position. Higher-level somatosensory areas receive input from the primary somatic sensory cortex, including the secondary somatic sensory cortex, located just below it, as well as Brodmann's areas 5 and 7b located just posterior to it. According to Kinsbourne (1995), there is a high-level body map in the left hemisphere. Neuropsychologists may be assessing this map when they ask a patient to name an indicated body part, or to point to a certain body part either on his own body or on a drawing. These areas and others work together to produce our conscious body image and to supply motor areas with the information necessary to execute complex movements. As with vision, a question arises about how the brain is able to merge the functions of several of these maps to produce a single, unified body image. For instance, if one somatotopic map signals pressure while another signals limb position, how is it that we are aware only of a single body that has both of these properties?

6.2   Theories of Anosognosia

Psychological Theories

For decades after its discovery, anosognosia was seen as primarily a psychiatric rather than a neurological problem. Weinstein and Kahn (1955), for instance, proposed that anosognosia was due to a nonconscious goal-directed reaction, or an adaptation against the stress caused by the illness. Others, however, believed that one cannot explain denial on sheerly psychological grounds, since that approach cannot account for the overwhelming tendency of these patients to have left-sided disabilities that are due to right hemisphere damage (Heilman et al. 1998). A purely psychological account also has trouble with the fact that there is a definite selectivity in the sorts of disabilities people deny. There are many case reports of patients who denied one disability while admitting others (Prigatano and Schacter 1991). One of Anton's original patients, for instance, denied blindness but was fully aware of and admitted her difficulty speaking (Anton 1899).

Representational Theories

In this view, the processes in the brain that represent the position, acceleration, force, and so on of the left arm are destroyed. Damage to areas known to be involved in representing the body (area 6) or its nearby space (area 8, frontal eye fields) are also known to produce neglect of the body and of space, respectively (Rizzolatti and Berti 1990). Another such deficit may occur in which the brain treats representations of intended movements as if they were representations of actual movements, something that might explain why these patients are so certain they moved when asked to.

However, most subjects with anosognosia do recognize their own hands, indicating that some representational ability is intact (Heilman et al. 1998). Similarly, Adair et al. (1997) reasoned that if the problem is that the left hemisphere is not receiving the necessary information, or not getting access to representations contained in the right hemisphere, there are ways to remedy this. Working with patients who had left-side paralysis caused by an injection that temporarily deactivated the right hemisphere, they used different approaches, such as having the patient read a number written on the left palm, and holding the left arm in the right field of vision, to get the left hemisphere to attend to the state of the left hand. This caused some of the patients to acknowledge paralysis (nine of thirty-two), but most (nineteen of thirty-two) continued to deny paralysis. There must be something more to the denial than a representation failure: "Whatever the neural basis for the anosognosic belief of having moved the paralyzed limb (we still do not know which premotor events take place in these circumstances), we must take note of the fact that this belief subsists independent of any feedback, even when the pathways that would carry kinesthetic information are spared and the patient can perceive passive movements" (Bisiach and Geminiani 1991, 36).

Emotional Theories

We will see in chapter 7 that the abilities of the isolated cerebral hemispheres to perceive and express emotions differ. Right hemisphere damage severely affects the person's emotional responses, and this may explain why anosognosic patients are so unconcerned about the damage (Gainotti 1972, 2001). Low autonomic responsiveness in a group of neglect patients supports this (Heilman et al. 1978), as does the finding, described in chapter 4, that the right hemisphere plays a dominant role in initiating skin-conductance responses. There is more on this in section 6.3, as well as in chapter 7.

Attentional Theories

According to these views, the patient's representations of space are intact; the problem is that he is unable to direct attention to the affected part of space. This is called the feedforward hypothesis for anosognosia (Heilman et al. 1998). To intend a movement is to set up an expectation. In this hypothesis, "weakness is detected when there is a mismatch between the expectancy of movement and the perception of movement" (Heilman et al. 1998, 1908). The patients do not set up these expectations; they do not intend to move, so they do not create an expectancy, a somatotopic motor representation of the intended action. Heilman et al. offered a thought experiment: "If your left arm were totally relaxed at your side, so that you did not attempt to move this arm, and a physician asked you whether your arm was weak, you would probably say, 'no, it is not weak'" (1998, 1907). They suggest that if we add to this situation that unknown to you, your left arm actually *was* paralyzed (e.g., by a beam of radiation) just before the doctor asked about it, you would be anosognosic. Why can't the patient tell he is not trying to lift his arm, however? This situation may be possible as described, but intuitively, anyone asked about the strength of his arm by a doctor would instinctively flex his left arm and look down at it. Patients do not do this, but the important question remains as to why they do not at least attempt to.

Disconnection Theories

According to Norman Geschwind (1965), denial occurs because the parts of the brain that produce verbal denial have been disconnected from the parts that contain the primary representations of the left side of the body and its space. Verbal denial is almost always generated by the left hemisphere, but crucial information about the body is normally contained in the right hemisphere's parietal cortex, a high-level association area (see section 6.3). "Right-sided association area lesions are more likely to lead to disability and to confabulatory response than those of the left side because the normally poorer linkage of the right side to the speech area, and possibly to other 'dominant' areas of the left hemisphere, is further weakened by the lesions" (Geschwind 1965, 600). The interesting notion, "confabulatory completion" was coined for cases in which a patient is shown an object that is half in his neglected visual field and half in his intact field, and who confabulates when asked about the color, shape, and so on of the part in the neglected half. Geschwind did not explain why the disconnection leads specifically to confabulation, however.

The reason an amputee with a phantom limb does not deny that his limb is gone is that higher-level association areas are intact. "Confabulatory denial of a lost limb is uncommon; although the patient frequently has a phantom limb, he nearly always treats it as a phantom. But denial of hemiplegia is common and probably more common in the presence of accompanying lesions of the association areas" (Geschwind 1965, 598). It is worth noting here that our patient with Capgras' syndrome, D.S., also had an amputated arm and at one stage denied that it was missing, perhaps a sign of a general failure to monitor and correct representations.

Geschwind allowed an additional, more general factor in confabulation, "the more demented or clouded the patient, the more likely he is to show confabulatory response in the presence of any defect of whatever origin" (1965, 597). Why don't the patients turn their heads to bring their left side into view, and why don't they notice and admit they are paralyzed? Note also that the study in which the paralyzed left arm is specifically brought to the attention of the left hemisphere works against a disconnection view (Adair et al. 1997). This failure could still be due to *intra*hemispheric disconnection in Geschwind's theory. As we will see in chapter 7, disconnection theories seem to gain support from the fact that split-brain patients also confabulate.

6.3   The Neuroscience of Denial

There is a basic problem with the technique of testing the claim that lesions in the right hemisphere are far more likely to cause unawareness of illness than lesions on the left: Damage to the left hemisphere may disrupt the patient's ability to speak, so that even though he might be unaware of his illness, he cannot issue verbal denials. Bisiach et al. (1986) argued that while patients with right hemisphere damage might verbally deny their disability, they should be able to express it through nonverbal means. However, as noted in chapter 1, Ramachandran (1995) gave patients with anosognosia a choice between a two-handed task, such as tying a shoe, and a simple one-handed task, such as screwing in a light bulb. The reward was greater for accomplishing the two-handed task, and patients chose, and failed at, the two-handed task. The issue of left-right asymmetry in neglect is less controversial; most experts agree that neglect is more severe and more frequent with right than with left hemisphere damage. Apparently the right hemisphere maintains representations of both sides of nearby space, while the left hemisphere represents mainly the right side of space (Mangun et al. 1994). This explains why right-sided neglect is so rare, although it is sometimes reported. Since the right side is represented in both left and right hemispheres, more severe damage is necessary to compromise both of these representations.

Role of the Parietal Cortex
Although damage to other locations can produce both neglect and denial, "the most frequent, profound, and enduring examples" (Cummings 1985, 50) of neglect occur with damage to right parietal regions (Critchley 1953). Lesions in the frontal eye fields, ventral area 6, the polysensory area of the superior temporal sulcus (Petrides and Iversen 1979), and the cingulate gyrus (Watson et al. 1978) can also produce neglect. It can be produced by lesions to certain subcortical areas as well, including the nigrostriatal system (which includes the caudate nucleus and putamen), the superior colliculus, the thalamic intralaminar nuclei (Rizzolatti et al. 2000, 504), and the lateral hypothalamus (although this may actually be caused by an interruption of nigrostriatal fibers crossing in the vicinity, according to Rizzolatti et al. 2000). Damage to some part of the inferior parietal lobe, as well as to the adjacent polysensory area of the superior temporal sulcus is necessary to cause neglect (Rizzolatti et al. 2000). In a review of cases, the most frequent type of damage in patients with asomatognosia (disowning of their arms) was a deep parietal lesion (Feinberg et al. 1990).

The parietal lobe can be divided into three sections: postcentral gyrus, superior parietal lobule (Brodmann's areas 5 and 7 in humans), and the inferior parietal lobule; these last two are sometimes referred to as the posterior parietal lobe. The inferior parietal lobe in humans consists of Brodmann's area 39—the angular gyrus, and Brodmann's area 40—the marginal gyrus (Rizzolatti et al. 2000). The parietal cortex is a high-level association area (Geschwind 1965) in that it receives extensively processed inputs from several different modality-specific areas. It also receives emotional-motivational input from the posterior cingulate gyrus (Mesulam et al. 1977). The parietal cortex (area PG in the rhesus monkey) is part of a large network for directed attention, along with the dorsolateral frontal area known as the frontal eye fields, but also including the superior temporal sulcus and the orbitofrontal cortex (Mesulam 1981). Damage to this network may account for the spatial disorientation frequently seen in confabulating patients, including those with ACoA aneurysm and Korsakoff's syndrome, and patients who deny disability. Perhaps when one of the component areas is disrupted, the whole network "goes down" for a period of time before it is able to compensate, readjust itself, and resume functioning without the destroyed area. "Some of the interconnected cortical areas . . . appear more critical than the others in the organization of directed attention. For example, unilateral neglect occurs after damage in the region of the [frontal eye fields], posterior parietal cortex, and the gyrus cinguli but not to the inferotemporal or orbitofrontal cortex" (Morecraft et al. 1993, 283). When the superior temporal sulcus was ablated on both sides, however, monkeys showed a pronounced inattention to visual, auditory, and somatosensory stimuli, which disappeared after two or three weeks (Petrides and Iversen 1979).

Recall from chapter 4 that the right inferior parietal cortex is one of the cortical areas capable of initiating a skin-conductance response. Activity in the right parietal cortex covaried significantly with SCRs when subjects performed a Bechara–Damasio-style gambling task (Critchley et al. 2000). When SCRs to an "uncomfortable but not painful" electrical stimulation were recorded in seven patients who had sustained a right hemisphere lesion and who exhibited flattened affect and neglect of the left side of space, five of these patients had no measurable skin-conductance responses at all (Heilman et al. 1978). "Emotional changes resembling those seen following frontal pathology may follow damage to other areas of the brain," including "right inferior parietal lobe pathology" (Stuss and Benson 1986, 137). Connections between the parietal lobes and the posterior cingulate cortex may be responsible for the emotional features associated with these parietal areas.

The right inferior parietal cortex is a high-level center of right hemisphere mental function. It represents our body situated in its nearby space. Its motivational-emotional inputs from the posterior cingulate may play a role in attaching a value to objects located in this space. The possibility of a valuation problem in neglect and denial knits together representational and emotional considerations. Even if the left arm is brought into view on the right side, unless the proper value and significance can be attached to it, the patient will ignore or even denigrate it. It cannot be his hand, he reasons, because he attaches no emotional significance to it. The damaged parietal areas may also be areas that attach a degree of significance, or value, to what they represent. The brain may capture the crucial difference in internal-external terms. The patient can no longer activate his left arm from the inside, as it were; now, he can only act on it from the outside.

6.4   Denial of Blindness

A sixty-eight-year-old stroke patient exhibited denial of blindness, or Anton's syndrome: "Behaviorally, he was unconcerned, disoriented to time and place, and failed to learn new material such as the name of the hospital, although he easily discussed older information. His pupils reacted to light, but he was unable to describe objects placed in front of him, count fingers, and so on. When asked to perform such tasks the patient often confabulated, giving an incorrect number of fingers, descriptions of items different from those held in front of him and so on" (Stuss and Benson 1986, 111). This patient made excuses that sound similar to those offered by hemiplegic denial patients: "He stated that he could not see the object or the fingers clearly. He strongly denied any problem with his vision, excusing his failure by saying that the lights were dim, that it was the dark of night, that he was not wearing the correct pair of glasses, etc" (1986, 111; notice that Stuss and Benson allowed that simply saying a number can count as a confabulation; there is more on this issue in chapter 8).

Many features of Anton's syndrome are similar to those seen in other confabulation syndromes, including frontal damage and memory problems. A patient who sat in front of an oncoming subway car in a suicide attempt sustained severe skull fractures. When asked in the hospital where she was, she stated that she was at her sister's house. "When confronted with evidence of the hospital surroundings, [she] replied: 'Oh, that peroxide and needles are my sister's. She has a skin problem.' She confabulated reasons as to why she made mistakes in determining how many fingers the doctor was holding up, saying on one occasion 'No, I can see it. I just have a headache.' She also blamed her failure on the lighting in the room and the angle of the bed" (McDaniel and McDaniel 1991, 102). There were signs of orbitofrontal damage; she also displayed "inappropriate, disinhibited behavior with evidence of poor insight and social judgement," as well as "verbal disinhibition with frequent threats ... toward nursing staff and paranoid ideation" (1991, 102). This woman was diagnosed with "an orbitofrontal lobe syndrome (disinhibited, impulsive behavior, inappropriate affect, emotional lability, irritability, poor judgement and insight, distractibility, and unconcern)" (1991, 103). The authors suggested that her Anton's syndrome was the result of the combination of frontal damage and damage to the visual system. Another patient was completely blind but claimed that he could see, during alcoholic withdrawal (Swartz and Brust 1984, 971). This patient admitted that some of the things he "saw" were hallucinations, but contended that he could nevertheless see in spite of this. "His total blindness was secondary to orbital trauma" (1984, 971).

In chapter 3 we saw that a strong case could be made for the claim that the amnesia and the confabulation in Korsakoff's and ACoA syndromes are caused by damage to two separate processes. Each type of damage can occur without the other, and the two have different time courses in many patients. Anton's syndrome seems to fit this pattern. There are two lesions—one that affects the patient's ability to see, and a second one that affects his ability to know whether or not he can see. In 1908, Redlich and Bonvicini made the interesting suggestion that Anton's syndrome is similar to Korsakoff's except that instead of not knowing that he has a poor memory, the patient does not know that he is blind. Anton's syndrome usually involves bilateral cerebral infarcts in areas fed by the posterior cerebral arteries (Heilman 1991, 55). The sixty-eight-year-old stroke patient, described earlier showed "bilateral posterior cerebral artery territory hypodensities" (Stuss and Benson 1986, 111). The damage can be produced either by bilateral lesions of the occipital cortex or the retrogeniculate visual radiations (Cummings 1985). It is interesting that the part of the circulatory system involved also serves the medial thalamus and Heilman reports another case of a patient with a mediodorsal thalamic lesion and memory problems (Heilman 1991).

The blindness in Anton's syndrome also may be caused by lesions at a wide variety of places along the visual pathway, so their particular location is not vital (Swartz and Brust 1984). What is necessary is a visual lesion and another frontal lesion that destroys the patient's ability to know that he cannot see. A patient who sustained frontal damage as well as bilateral traumatic optic neuropathy as a result of a car crash would sometimes admit blindness, but also insisted that he could see normally, with the proper illumination (Stuss and Benson 1986). One suspects that optic neuropathy caused the blindness, while frontal damage caused an inability to know about the blindness. Anton's syndrome is also similar to Korsakoff's in that it often involves memory problems. Some patients have memory problems and are in a confusional state (Hecaen and Albert 1978). They frequently have disorders of recent memory, and confabulate about recent memories (McDaniel and McDaniel 1991).

6.5   Anosognosia and the Other Confabulation Syndromes

The frequent claim that confabulation involves two problems, for example a memory problem, and a monitoring problem, seems to run aground in the case of anosognosia caused by damage to a single area, especially one not in the frontal lobes, such as the right inferior parietal cortex. It seems understandable that damage to a body-representing area such as the right inferior parietal might lead to unawareness about the body, but how on earth could it cause the patient to deny that the left side of his body is his? "An objection to the effect that such disorders might be due to a local impairment of an analogue wired-in device, subserving cognitive processes but not itself endowed with authentic cognitive competence is contradicted by the fact that no superordinate cognitive mechanism detects, and compensates for, these disorders" (Bisiach and Geminiani 1991, 30). Where are the frontal checking processes?

The existence of local confabulation of the type that is seen in anosognosia is awkward for theorists of confabulation who posit two factors. Somehow, something must go wrong with a special knowledge domain, and the second phase has to be debilitated in exactly that domain. If asking for two lesions was too much, asking for the second lesion to be this specific seems to be unreasonable. One possibility is that there is a superordinate cognitive mechanism or several, but that they are at least temporarily compromised by the parietal damage.

Another way that a single parietal lesion could have double effects is if it plays a role in both the first and second phases of confabulation. The right parietal cortex may play a role both in generating thoughts about the state of the body and in checking claims about the state of the body. Damage to it will then produce both types of damage necessary for confabulation. First, a false claim will be generated because the area responsible for that knowledge domain is damaged and some other much less competent area has generated that answer. Then the answer cannot be properly checked because the only area that can do this is damaged. The fact that the right parietal cortex is not a frontal area does not mean it is not a high-level cortical area. It also makes sense that higher levels of processing in the right hemisphere might be more closely related to perception than to action, given its strong role in perception.

In their penetrating discussion of anosognosia, Bisiach and Geminiani noted that the system that produces the confabulation, which they called simply "language," is strangely insensitive to the problem. "Language, by itself, is unable to compensate for the representational lacuna that hemineglect patients may reveal in the mental descriptions of familiar surroundings. It is passively caught in the tangle of spatially circumscribed misrepresentation and wild reasoning of anosognosic patients. This fact suggests that it is not an autonomous representational system but a means for communicating contents supplied by nonlinguistic thought processes occurring on the basis of an analogue representation medium" (1991, 30–31). The language system creates the confabulation, but there is no reason to think that it also is the system that should be able to detect problems in its verbal creations. The phenomenon of confabulation in general attests strongly to this.

Three processes are under discussion here: (1) processes centered in the inferior parietal and frontal eye fields that represent the body and its nearby space, (2) processes that create the confabulation (presumably in the left hemisphere), and (3) frontal processes capable of checking the confabulation and finding it wanting. What is puzzling is not so much why the linguistic processes cannot detect the error in what they have created, but why the frontal processes cannot detect it. Again, an explanation for these phenomena is that the system centered on the right inferior parietal is vital to the checking-monitoring processes centered in the orbitofrontal lobes, so that the parietal damage at least temporarily disables the checking processes. As both Geschwind, and Bisiach and Geminiani noted, "damage to lower levels . . . abolishes those earliest states of sensory awareness . . . but not the nervous system's ability to monitor the lack of awareness in the form of a sensory scotoma. Damage to higher levels, in contrast, abolishes this ability, and the patient is unaware of his hemianopia or hemiplegia" (Bisiach and Geminiani 1991, 32). It may indeed be true that there is no higher-level process in the brain that can detect the types of phenomena the inferior parietal cortex detects, but this still has circumscribed implications about the nature of all higher-level brain processes.

There may not be a single, grand frontal process that monitors everything, but this does not mean there are not several processes that do this. The fact that several of them exist may explain why confabulation can be restricted to certain domains. Anosognosia seems to involve clear cases of confabulation restricted to certain domains, such as knowledge of the left side of the body. "The view that delusions that can follow brain injury do not result from global changes in mental function is strengthened by important observations in cases of anosognosia which demonstrate that patients with more than one neurological deficit may be unaware of one impairment yet perfectly well aware of others" (Young 2000, 51). It is still also possible that what is damaged is more global than it appears, particularly if it plays a role in mind-reading processes (more on this later). Problems in more global processes might appear local as long as the other knowledge domains are able to generate largely correct claims, so that no frontal process needs to inhibit, filter, or correct them. The problem in more global processes might show up in overgeneration (or false positives) in other domains, given the right testing situations.

One fascinating suggestion is that hypochondria is a mirror-image syndrome of anosognosia (Heilman 1991). The worries of the hypochondriac are the opposite of the serene unconcern of the anosognosic. The hypochondriac believes he or she is ill and is not, whereas the anosognosic is ill, but does not believe it. Evidence suggests that hypochondria and obsessive-compulsive disorder often occur together (Neziroglu et al. 2000), as we would expect if they are both at the opposite end of a continuum from confabulation.

Unawareness and Consciousness

One thing of great interest in unawareness phenomena is that they may be able to shed light on the connection between confabulation and consciousness. Why do patients not notice that the left side is absent from consciousness? In one view, conscious states span several cortical areas, which are temporarily integrated into a single "field," a view referred to as the integrated field theory of consciousness (Kinsbourne 1988, 1993). In neglect, in this view, representations of the body are not sufficiently developed to be integrated into the person's larger conscious states. The question remains, though, why we do not notice that they have not been integrated into the current conscious state. Here is one possibility: According to the theory, the set of cortical areas currently bound into the current state is constantly changing. If something in consciousness were always signaling a gap from areas that have not been integrated, this would most likely seriously disrupt the functions associated with consciousness. Hence, in the absence of a positive error signal, we assume that our conscious states are complete.

An augmented version of the disconnection theory that is specifically directed at questions about consciousness postulates the existence of a conscious awareness system (CAS) that normally interacts with several modular processes, including one responsible for representing space (Schacter 1990): "Under normal conditions, CAS takes as input the highly activated outputs of various modules; weakly activated outputs do not gain access to the awareness system. Suppose that conditions exist under which CAS can be selectively disconnected from a damaged module. The awareness system would then no longer receive those highly activated outputs that 'alert' it to the state of that module's activity. Thus, with respect to CAS, the disconnected module would be in [a] perpetual state of resting or baseline activity, no different from any other normally functioning module that is in a low state of activation" (Schacter 1990, 174). "Unawareness of deficit could result from such a disconnection" (1990, 174). An executive system also was posited that has access to the CAS and performs checking-monitoring functions: "In the model, the executive system is held to be responsible for intentional retrieval and ongoing monitoring of complex internal activities" (Schacter 1990, 173). Why, though, can disconnection alone cause unawareness, and why doesn't the executive system detect the problem?

Other writers traced the patient's failure to notice a difference in what he is consciously aware of to his inability to tell where certain representations originated. Patients with Anton's syndrome often have visual hallucinations (Heilman 1991, 55). An interesting theory of the nature of the problem is based on the evidence that vision and visual imagery share certain brain processes: "When a subject is asked to use visual imagery he often disengages from focusing visual attention and may even close his eyes to reduce visual input. Psychological studies have demonstrated that visual imagery and visual processing may be mutually inhibitory. This mutual inhibition suggests that both vision and visual imagery may compete for representations on a visual buffer" (Heilman 1991, 57). In this account, the syndrome occurs because visual imagery is mistaken for actual vision: "Patients with cortical blindness may still have visual imagery, and it is the visual imagery in the absence of visual input that may not only induce visual hallucinations but also may lead to anosognosia. It is possible that the speech area or the monitor receives information from the visual buffer rather than visual input systems. Because the buffer contains visual representations that come from imagery the confabulatory response may be based on this imagery" (Heilman 1991, 57). It may be that the damage to frontal processes often reported in Anton's syndrome accounts for the patient's inability to distinguish hallucinations from vision.

Anosognosia and Mind Reading

Is there a connection between anosognosia and theory of mind? One large study noted that anosognosic patients had deficits in recognizing facial emotions as well as emotional tonality in voices (Starkstein et al. 1992). Patients with neglect and denial also confabulate about whether or not *other* patients moved their arms (Ramachandran and Rogers-Ramachandran 1996). Recall from chapter 5 that damage to the right inferior parietal cortex can produce a deficit in recognizing facial expressions (Adolphs et al. 1996). In a subsequent study, these authors found activity in the right parietal cortex during a mind-reading task and suggested that the somatotopic areas may be producing a simulation of the seen face. Denial patients also seem to show failure in what we might call emotion sharing. We are set up to register emotions similar to those of people we are with, perhaps to synchronize our emotional responses. What the relatives of denial patients find so alarming is the way that the patients fail to share their concern. Normally this sort of concern is quite infectious.

The right hemisphere's inferior parietal cortex contains an interesting representation system centered around a system of body representations.

Moving out, other systems represent one's nearby personal space. What we have said thus far is true of the dorsal visual stream. However, my nearby personal space contains objects, some of which I am interested in, others that I have no interest in. Those objects are represented primarily in the ventral visual streams running through the temporal lobes. Our awareness of important objects in our nearby space must thus be embodied in a combination of the ventral and dorsal streams. In this respect, it may be significant that the superior temporal sulcus, an important area of damage in anosognosia, receives input from both "what" and "where" visual streams (Heilman et al. 2000). This conception makes sense in the case of objects that are within arm's reach, but our mental lives go beyond this; we think about people and things that are far distant from us in both space and time. We know how to access those things and people, though, and we know how they can access us. As I noted in chapter 5, this representation system may have a self-mode in which it functions to represent our personal space, and a mind-reading mode in which it functions to represent other people in a spatial milieu.

Damaged parietal areas may also be areas that attach an intrinsic significance or value to what they represent. I noted earlier that if the same areas that represent our own bodies also sometimes function to represent the body of the person we are perceiving, this can explain how humans attach value to each other. One interesting piece of evidence for the idea that the right inferior parietal can represent nonactual space is that "out-of-body experiences" could be induced by applying current to electrodes embedded in the right angular gyrus of a patient with epilepsy (Blanke et al. 2000). It is also of interest that other damage areas involved in producing neglect are important for mind reading, including areas around the superior temporal sulcus (Rizzolatti et al. 2000) and the ventral motor areas found to contain mirror neurons.

6.6 Conclusion

One of the most confounding facts to anyone seeking a theory of the neural locus of confabulation is the way that anosognosia can be caused by sites distant from the frontal areas associated with confabulation in other syndromes, such as Korsakoff's syndrome and ACoA aneurysm. As we saw in earlier chapters, evidence from those disorders points toward the orbitofrontal lobes as a crucial damage or disconnection site in confabulation. Although the orbitofrontal lobes and inferior parietal cortex are spatially distant, they are both part of the ventral cortical moiety (Pandya and Yeterian 2001), and they both participate in certain mind-reading circuits, as well as the network for directed attention (Mesulam 1981).

The inferior parietal cortex relates to the claims and findings of the previous chapters in two ways:

1.   The right inferior parietal cortex may be part of a mind-reading circuit. The inferior parietal cortex is strongly interconnected with the same temporal lobe area that the orbitofrontal cortex is connected with, that is, areas in and around the superior temporal sulcus. Recall that these areas contain neurons that respond most strongly to faces. In chapter 5 I explored the connection between confabulation and the ability to understand the minds of other people, and proposed that the connection between the orbitofrontal cortex and temporal lobe areas might be important both for mind reading and in the genesis of the strange confabulations of patients with misidentification disorders such as Capgras' syndrome.
2.   The inferior parietal cortex plays an important role in the generation of SCRs. We also saw in chapter 4 that there are interesting connections among confabulation, lying, sociopathy, and the SCR. Damage to the emotional functions of the inferior parietal may explain why these patients are so blasé about their problems, and also why they may disown their left limbs: Even when they see the limb, it is of no value to them.

Other reported connections have been reported between anosognosia and the other confabulation syndromes. Misidentification syndromes frequently cooccur with anosognosia (Mentis et al. 1995). Valuable comparisons were also made between denial patients and the lobotomy patients we examined in chapter 4: "Our findings are related to some of the problems of prefrontal lobotomy. This procedure when performed bilaterally evidently creates a milieu of brain function sufficient for the existence of anosognosia. After the operation, patients may show all aspects of the syndrome of denial: denial of the operation, disorientation, reduplication, confabulation, paraphrasic language, mood changes, and urinary incontinence" (Weinstein and Kahn 1955, 126).

In the next chapter we look at the final type of confabulator: the split-brain patient. These individuals share with many anosognosic patients the inability to report stimuli on their left sides. Also, some of the theories we examined here have been applied to these patients, such as Geschwind's disconnection theory. Both anosognosic patients and split-brain patients may be confabulating as a result of a lack of information combined with an inability to notice this lack.

7 The Two Brains

Now, if lack of harmony in the exterior sensory system upsets the brain's perception, why should not the soul perceive confusedly, when the two hemispheres, unequal in strength, fail to blend the double impression that they receive into a single one.

—Francois Bichat (1805)

7.1 Confabulations by Split-Brain Patients

The first split-brain operations on humans took place in the early 1940s, when Van Wagenen and Herren (1940) performed the surgery in an attempt to prevent the spread of epileptic seizures from one hemisphere to the other. These patients showed no real improvement, so the surgeries were stopped for the next twenty years (Funnell et al. 2000). In the early 1960s, however, Bogen and Vogel (1962) speculated that the early attempt had failed because not all of the three commissures had been removed. They tried this on a patient with severe, life-threatening seizures and it succeeded in controlling them. Since then, increased knowledge about the functional connections of the different commissures has allowed the operation to be refined so that typically only a portion of a commissure is removed. Two classic types of split-brain operations are performed. In a *callosotomy*, both the corpus callosum and the hippocampal commissure are removed. In *commisurotomy*, the corpus callosum, hippocampal commissure, and anterior commissure are all sectioned; section 7.3 examines the anatomy of the commissures.

When Roger Sperry, Michael Gazzaniga, and others began testing split-brain patients after surgery in the early 1960s, they noticed that the left brain tended to confabulate about actions performed by the left hand, which is controlled by the *right* brain. They found that one way to produce confabulation in a split-brain patient is to lateralize a stimulus to the right hemisphere, then ask the left hemisphere a question about the stimulus or the (right hemisphere-generated) response. Since the optic chiasm is left intact by the split-brain operation, the subject must fixate on a central focal point while images are flashed to the left visual field. The images are flashed just long enough to allow a hemisphere to detect them, but not long enough to allow the patient to shift his or her eyes away from the focal point, which could allow the other hemisphere to detect the stimulus. Various lenses and goggles have been developed for lateralizing visual stimuli without requiring voluntary fixation, but as long as the image remains on the proper side of the fixation point, it should be successfully lateralized to the contralateral hemisphere.

In one experiment that produced confabulations, Gazzaniga and his colleagues flashed commands to the right hemisphere, which was quite capable of obeying them. But when the patient was asked why he performed the action, the left hemisphere would confabulate: "When P.S. was asked, 'Why are you doing that?' the verbal system of the left hemisphere was faced with the cognitive problem of explaining a discrete overt movement carried out for reasons truly unknown to it. In trial after trial, when queried, the left hemisphere proved extremely adept at immediately attributing cause to the action" (Gazzaniga et al. 1977, 1146). The patient's adeptness makes one wonder whether people might be doing this often in their everyday lives. As we will see, it makes some writers wonder whether we might all be doing something like this. Gazzaniga's examples give the impression of a left hemisphere hypothesizing after the fact about the reason or motive for an action: "When 'laugh,' for example, was flashed to the right hemisphere, the subject commenced laughing, and when asked why, said 'Oh, you guys are too much.' When the command 'rub' was flashed, the subject rubbed the back of his head with his left hand. When asked what the command was, he said 'itch'" (Gazzaniga et al. 1977, 1146). When the command is 'walk', "a subject typically pushes her chair back from the testing table and starts to walk away. You ask, 'Why are you doing that?' The subject replies, 'Oh, I need to get a drink'" (Gazzaniga 1998, 133).

Gazzaniga argues that split-brain patients also confabulate about the cause of their emotional responses. In one experiment a violent scene—a person throwing another person into a fire—was lateralized to the right hemisphere of a split-brain patient. When the patient was asked what she saw, she responded: "I don't really know what I saw. I think just a white flash. Maybe some trees, red trees like in the fall. I don't know why, but I feel kind of scared. I feel jumpy. I don't like this room, or maybe it's you getting me nervous" (Gazzaniga 1992, 126). According to Gazzaniga, the right hemisphere first perceived the violent act. This then caused the brain's emotional system to produce a negative emotion, which the left hemisphere detected because the emotional system has significant subcortical components that span the two hemispheres. The left hemisphere then produced a confabulated reason for the emotion, typically by pointing to features of the patient's surroundings.

Such lateral theories of confabulation typically attribute it to a problem in the interaction between the two cerebral hemispheres. Until recently, there has been little communication between the frontal and lateral confabulation theorists, and neither side has endeavored to extend its theory to cover the other's phenomena. This chapter starts with an examination of the history of these lateral theories, beginning with Geschwind's and moving to Joseph's, but it focuses heavily on Gazzaniga's. We then discuss ways in which confabulation in split-brain patients relates to the other confabulation syndromes we have examined.

One conclusion several sources point to is that the orbitofrontal cortices of the left and right hemispheres are highly integrated and often function in tandem. This would explain why damage to either of them or to their commissural connections might have similar effects. We have seen several clues thus far pointing to some sort of lateral dynamic in confabulation: Disinhibition occurs more frequently with right frontal lesions; mind-reading processes seem to be based more in the right hemisphere than in the left; the preponderance of misidentification syndromes involve right hemisphere damage; neglect with denial occurs predominantly with right hemisphere lesions; and autonomic responses such as the skin-conductance response are initiated primarily by the right hemisphere (Zoccolotti et al. 1982). This chapter focuses on elucidating the physical-causal connections between the lateral and frontal phenomena. In the next chapter I contend that confabulation by patients with both lateral and frontal damage can be unified into a single concept according to which confabulation is due to knowledge deficits.

7.2 Hemispheric Differences

It is immediately apparent upon looking at the brain that there are two brains atop the brainstem: two cerebral hemispheres, two thalami, two amygdalas, and so on. So pervasive is this duality that when Descartes examined the brain, looking for the place where mind and body come together, he chose the pineal gland because it was the only organ he could find that was not part of a pair, and he assumed that our conscious lives are unified—single, not dual. The exact reason for this wholesale duality is a subject of debate, but it begins far back in our evolutionary past. Even more curious is the way that the brain likes to cross its connections with the body, so that the left side of the body is represented and controlled more by the right hemisphere and vice versa. These crossed connections between brain and body are referred to as contralateral, as opposed to connections on the same side, between for instance, the left hemisphere and the left side of the body, which are called ipsilateral.

It was initially thought that the human brain was unique in assigning specialized functions to the two hemispheres, and that animal brains are functionally symmetrical. As more and more evidence of asymmetry of function in the animal brain appeared, however, this latter claim was revised (see Hellige 1993). It may nevertheless remain true that the human brain is more asymmetrical than the brains of animals. This is part of a general cross-species trend; the more complicated a species' brain is, the more restricted its interhemispheric connections (Pandya and Seltzer 1986), allowing the two hemispheres to operate with greater independence. Hemispheric specialization, and the resulting asymmetry, has the potential to greatly increase both the representational space available and the computing power of a brain, although it may require a more complex executive system to coordinate activity and to produce goals and behavior that are unified.

Sometime after the first split-brain operations, in the late 1960s a popular movement began that assigned different characteristics to the two hemispheres. The left hemisphere was typically claimed to be analytical and logical, and the right was supposed to be more "intuitive," artistic, and creative. People were urged to put their right hemispheres to better use, by drawing with the left hand, for instance. Beginning in the 1970s, the inevitable backlash came: orthodox science showed that this story about the different hemispheres was just that, a story. As with any backlash, however, it went too far, to the point where well-informed people were ready to argue against just about any mental function being exclusively lateralized to one hemisphere or the other.

Although the simple division of functions posited by popular writers proved to be mistaken, several laterality findings are now well supported: Language abilities are strongly lateralized to the left hemisphere, especially in right-handed people, but also in most left-handed people. Despite reports that the right hemisphere is capable of language, many researchers are pessimistic about its ability, even given time and training, to match the capacity for abstraction and explanation shown by the left hemisphere (see Gazzaniga 1983; Zaidel 1976, 1983). Researchers have found that the right hemisphere is better at understanding than producing language, and better at understanding nouns than verbs. It also specializes in producing and analyzing the emotional tone of speech. It is chilling to hear the monotone of a patient with *aprosodia*, a condition that is due to right hemisphere damage in which all emotional tonality is gone from the patient's speech. Sometimes these patients are also unable to interpret the emotional tonality in the speech of others. Whereas both hemispheres can initiate spontaneous facial expressions of emotion, only the left can voluntarily produce facial expressions (Gazzaniga 1995b).

Memory findings are also noteworthy. Split-brain patients have a memory deficit on free recall tasks; for instance, a task in which a patient is read a list of twenty words and asked later to recall as many as he can. Only patients who have had the posterior portion of the corpus callosum removed show this effect (Gazzaniga 1995b); those with an anterior section have normal recall. Jha et al (1997) found that callosotomy patients could remember verbal material at a normal rate, but performed poorly at remembering pictorial material. They interpreted this to mean that memory for pictures requires interhemispheric interaction. The left hemisphere has better recall for verbal material, whereas the right hemisphere has better memory for faces (Miller and Gazzaniga 1998). A still-robust theory suggests that the two hemispheres play different roles in episodic memory: The left hemisphere is responsible primarily for encoding items in memory, whereas the right focuses on retrieval (Tulving et al. 1994). Known as the HERA (hemispheric encoding and retrieval asymmetry) model of memory, it has received support from several imaging studies showing right hemisphere prefrontal activity in subjects who are retrieving memories (Nyberg et al. 1996). Marcia Johnson and her colleagues (1996) have suggested that this right frontal activity represents what they refer to as heuristic memory processing, such as those involved in checking source memory attributions.

Someone must lead the dance; the human brain is designed so that the left hemisphere does the leading. Luria (1966) argued that the development of language in the left hemisphere greatly increased its control capacities. Gainotti (2001) has suggested that the right hemisphere is specialized for automatic functions of the spatial orienting system, whereas the left hemisphere achieves more conscious, intentional spatial orienting. It may be that the need for there to be one hemisphere (i.e., the left) with predominant connections to the motor areas forced the right hemisphere into a fundamentally inhibitory mode. Starkstein and Robinson (1987) note that most patients with a disinhibition syndrome have a right hemisphere lesion. Since it is unable to initiate voluntary actions, the right hemisphere's main role may be to verify and possibly inhibit the action intentions of the left hemisphere. The situation is rather like U.S. politics; the party that wins the presidency becomes the primary planning and executing body, whereas the other party can only inhibit or vote down these actions. For pragmatic purposes, it would also be good for the left hemisphere to be, in a sense, self-sufficient. It has to be ready to speak for the entirety of the brain, whether or not the right hemisphere can help it by supplying, or allowing access to, information.

We saw in chapter 2 that the brain often lets opposing sets of processes balance and temper one another. There are several hypotheses that the two hemispheres balance and oppose each other in different ways. One theory is that each cortical area, when working, is set up to inhibit its corresponding ("homologous" or "homotopic") area in the other hemisphere through the corpus callosum (Cook 1984). A similar suggestion is that the two hemispheres hold each other in balance, sometimes by inhibiting each other (Kinsbourne 1975). This hypothesis involves less specific effects than the first; it proposes that the hemispheres inhibit each other's general level of activity, rather than inhibiting selected cortical areas.

7.3   Anatomy of the Cerebral Commissures

The left and right hemispheres are connected by three bands of fibers. The largest of these by far, the corpus callosum (figure 7.1), contains more than 200 million nerve fibers (Funnell et al. 2000). It connects areas of one hemisphere's cortex with homologous cortical areas in the other hemisphere. The hippocampus also has its own commissure, although it is smaller in humans than in animals. The anterior commissure is a small separate link that looks rather like an evolutionary afterthought, connecting parts of the temporal and orbitofrontal lobes.

Corpus Callosum

The corpus callosum generally allows two-way communication between homotopic areas of the two hemispheres. It does so selectively, however; some regions of the cortex have dense connections to the other hemisphere and others have none (Pandya and Seltzer 1986). Pandya and Seltzer also note that somatosensory and motor areas that represent central parts of the body such as the torso have extensive interhemispheric connections, whereas areas representing the more distal body parts, such as the hands, tend to have few such connections. This may be relevant for anosognosia for hemiplegia, since it occurs with paralysis of the most lateral parts of the body, and not with more central paralysis.

Anterior Commissure

The anterior commissure may be significant for our project of understanding confabulation, since it interconnects areas such as the orbitofrontal cortex and parts of the temporal lobes. We also saw in chapter 3 that it lies in the area of distribution of the anterior communicating artery. The anterior commissure is thought to connect much wider territories in humans than in nonhuman primates (DiVirgilio et al. 1999). Demeter et al. (1990) found that the anterior commissure receives input from the entire temporal lobe (primarily from its anterior portions), as well as area 13 of the orbitofrontal cortex (see figure 4.3) and the amygdala. Their diagrams show that injections in the portions of the anterior and inferior temporal lobes that contain face-responsive cells contribute strongly to the anterior commissure (Demeter et al. 1990). The corpus callosum itself primarily connects the posterior portions of the temporal lobes, leaving the interhemispheric connections between the more anterior and inferior temporal lobes, as well as those between the left and right orbitofrontal cortices, to the anterior commissure. Demeter et al. also argue that the connection from the orbitofrontal cortex to the entorhinal cortex in the other hemisphere is contained in the anterior commissure. Since this is a connection between the

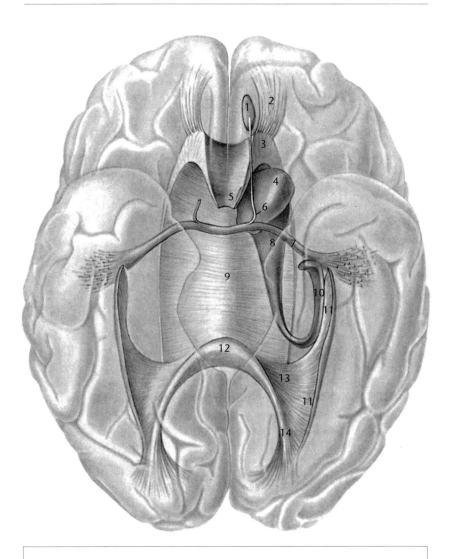

1 Bulbus olfactorius
2 Forceps minor
3 Genu corporis callosi
4 Caput nuclei caudati
5 Rostrum corporis callosi
6 Commissura anterior, crus anterius
7 Commissura anterior, crus posterius
8   Corpus nuclei caudati
9   Truncus corporis callosi
10 Cauda nuclei caudati
11 Tapetum
12 Splenium corporis callosi
13 Radiatio corporis callosi
14 Forceps major

Figure 7.1
The corpus callosum and the anterior commissure. (From Nieuwenhuys et al.,
1988, with permission.)

orbitofrontal cortex and a memory area, it may be important for memory confabulations.

Hippocampal Commissure
This commissure was removed in both types of split-brain operation, commissurotomy and callosotomy, primarily because researchers strongly suspected it was responsible for transferring seizure activity from one hemisphere to the other. Lower animals have both a ventral and a dorsal hippocampal commissure, but the ventral hippocampal commissure gradually disappears in higher primates and is often not present in human brains (Gloor et al. 1993). The human dorsal hippocampal commissure is very much present, however. It crosses the midline just below the rear portion of the corpus callosum. It contains fibers that originate in the hippocampus (in the presubiculum). Several areas of the temporal lobes (TF, TL, and TH) also enter the hippocampal commissure (Gloor et al. 1993). These fibers briefly become part of the fornix, then cross over to join their (typically homologous) areas in the other hemisphere.

7.4   Lateral Theories of Confabulation

Geschwind's Disconnection Theory
In 1965, neurologist Norman Geschwind laid out a detailed theory of confabulation that he applied not only (as we saw in the previous chapter) to anosognosia caused by a right inferior parietal lesion, but also to split-brain patients. He noticed that sometimes we are aware of perceptual deficits and at other times we are not. The difference, he stated, is that we are aware of the damage when it occurs in a perceptual area, but are not aware when the damage is in a higher-level association area. "Confabulation is much more likely in the presence of disease of association cortex or association fibres (either commissural or intrahemispheric) than it is in cases of damage of the primary sensory pathways up to and including the primary sensory cortex" (Geschwind 1965, 597). The right inferior parietal cortex is one such association area, so named because it receives input from several different sensory modalities. When claims generated by the left hemisphere fail to reflect information contained in the right hemisphere, this is attributable to disconnection.

Geschwind attributed the problem to a peculiarity in the communication between the association area and the speech area: "I can propose a highly hypothetical mechanism.... Since the speech area has in the normal no direct contact with the visual cortex proper, destruction of the calcarine [visual] cortex leaves the speech area still innervated by visual association areas. The association areas in this case receive no stimulus from the visual cortex and send the 'message' to the speech area that there is no visual 'message,' i.e. that all is black. This presupposes that the association areas never fail to send a 'message' to the speech area and that they always send *positive* messages of light or darkness." When processes the next level up, the association areas themselves, are destroyed, however, the message of light or darkness cannot be sent: "The destruction of the association cortex or fibres corresponds to a totally unphysiological state—one in which *no* message is being received by the speech area. Since this is not interpreted (not being one of the normal states) the speech area begins to react to random messages received from subcortical pathways or indeed responds to its own spontaneous firing." Sometimes the break is not complete, and some information gets through: "[The speech area] may respond to incomplete information coming over subcortical pathways which are inadequately extensive to carry all the information about the seen object; this may lead to errors which are less bizarre than those in which no information at all gets to the speech cortex" (Geschwind 1965, 599).

In a 1986 article, Rhawn Joseph proposed several new twists to the disconnection theory. Referring both to Korsakoff's confabulations and anosognosia, he said that "in many instances these behaviors, denials, and conflicts are the result of the differential organization of the right and left cerebral hemispheres (as well as that of the limbic system) and their inability to completely communicate and engage in information transfer and exchange" (1986, 517). Having two specialized hemispheres creates a special problem of reintegrating the results of different types of processes performed on different types of representations: "The right and left cerebral hemispheres not only speak different languages (insofar as visual–spatial, nonlinguistic, social–emotional, sequential and other processes are concerned), they process and express select forms of information that often cannot even be recognized or detected by the other half of the brain" (1986, 517).

Joseph calls the process that actually produces the confabulation "the language axis." As was the case with Bisiach and Geminiani, Joseph emphasizes the modularity and limited nature of the language axis: "It must be stressed that the aspect of the individual that confabulates and reports these erroneous or absurd statements is the language axis (i.e., Broca's and Wernicke's areas, the inferior parietal lobule, thalamus....), a multi-channeled neuronal network that is, in itself, without judgment, morals, or, when confronted in isolation, comprehension. Rather, this dynamic interconnected neuronal linguistic network acts only to assimilate what is available to it, advancing and organizing information that is received from other regions of the brain" (1986, 516–517). He attributed this blindness of the language axis to a division of cognitive labor: "Because the language axis is dependent upon these other brain areas to analyze and determine information relevance, it cannot, in itself, realize the absurdity of the positions it attempts to advance *even when the brain is undamaged"* (1986, 517). But again, why don't frontal processes detect the problem? The answer may be that the split-brain operation disables frontal processes because of their bilateral, integrated nature.

Gazzaniga's Theory

Michael Gazzaniga's theory of confabulation is in the tradition of Geschwind and Joseph. To explain the confabulations that split-brain patients engage in, Gazzaniga (1995a, b) proposed that the left hemisphere contains an *interpreter*, a cognitive module that functions to produce a verbal explanation of the agent's activities, based on the sources of data available to it. "The very same research that exposed shocking differences between the two hemispheres also revealed that the left hemisphere contains the interpreter, whose job it is to interpret our behavior and our responses, whether cognitive or emotional, to environmental challenges" (1998, 174). Confabulation occurs when the interpreter concocts an explanation without benefit of vital knowledge possessed only by the right hemisphere. Gazzaniga does not say where exactly in the left hemisphere the interpreter is, although he does suggest that it lies in the distribution of the middle cerebral artery.

Gazzaniga applied his theory to reduplicative paramnesia: "This patient has a perfectly fine interpreter working away trying to make sense out of what she knows and feels and does. Because of her lesion, the part of the brain that represents locality is overactive and sending out an erroneous message about her location. The interpreter is only as good as the information it receives, and in this instance it is getting a wacky piece of information. Yet the interpreter still has to field questions and make sense out of other information coming in that is self-evident. As a result, it creates a lot of imaginative stories" (1998, 136–137). When he says, "the interpreter is only as good as the information it receives," this seems to mean that the interpreter itself has no capacity to verify its input. This function may fall to the frontal processes described in chapters 3 through 7, as well as the function of the evaluating the output of the interpreter.

Part of the interpreter's problem seems to be that it has access only to the final products of the systems that provide it with input. "The functioning modules ... operate largely outside the realm of awareness and announce their computational products to various executive systems that result in behavior or cognitive states. Catching up with all this parallel and constant activity seems to be a function of the left hemisphere's interpreter module" (2000, 1394). Again, though, this covers only the first phase of confabulation. It refers to the creation of the confabulation, but not to the patient's inability to realize it is false. The way for this sort of function to be accomplished given the resources Gazzaniga mentions would be to feed the interpreter's explanation back into consciousness. Just as with the other processes, it is announcing its computational products to various executive systems. As we saw in chapter 4, some of these executive systems can examine and reject unsatisfactory explanations. Just as with other modules, the interpreter creates its explanation "outside the realm of awareness" and announces it to awareness.

The interpreter does not merely explain; it constructs stories about events in time. The interpreter "is tied to our general capacity to see how contiguous events relate to one another" (Gazzaniga 1998, 26). Gazzaniga attributes to the interpreter self-representing and self-creating functions similar to those we saw in chapter 1 that Dennett said are part of human nature. "The interpreter constantly establishes a running narrative of our actions, emotions, thoughts, and dreams. It is the glue that unifies our story and creates our sense of being a whole, rational agent. It brings to our bag of individual instincts the illusion that we are something other than what we are. It builds our theories about our own life, and these narratives of our past behavior pervade our awareness" (1998, 174). Gazzaniga posits an interesting additional function for the interpreter; it "allows for the formation of beliefs, which in turn are mental constructs that free us from simply responding to stimulus-response aspects of everyday life" (1995b, 1394). As we have seen, however, the production of thoughts is only part of the belief formation process. A thought must also pass certain checks before it is allowed to become a belief, in the normal brain.

False memories are of interest to us because they bring the interpreter together with the autobiographical memory system in Gazzaniga's approach: "The interpreter and the memory system meet in these false recollections. As we spin our tale, calling on the large items for the schema of our past memory, we simply drag into the account likely details that could have been part of the experience. Such easily disprovable fabrications go on all the time" (1998, 142). Recall from chapter 3 that the extra work the left hemisphere has to do in creating false memories may have been captured by brain-imaging techniques (Schacter et al. 1996b; Cabeza et al. 2001). Gazzaniga employed a version of Geschwind's theory when he discussed anosognosia: "When a neurologist holds this patient's left hand up to his face, the patient gives a very reasonable response: 'That's not my hand, pal.' His interpreter, which is intact and working, can't get news from the parietal lobe because the lesion has disrupted the flow of information" (1998, 136).

Gazzaniga assigns several high-level tasks to the interpreter: it "fields questions"; it "makes sense out of [incoming] information" (1998, 137); it "interpret[s] our behavior and our responses, whether cognitive or emotional, to environmental challenges"; it "constantly establishes a running narrative of our actions, emotions, thoughts, and dreams" (1998, 174).

Often statements such as these elicit the charge of homunculus fallacy: the claim that one is attempting to explain a brain function by positing something that has all the capabilities of a full-fledged human being, and hence not explaining anything at all. All that is necessary to avoid this charge, however, it that the theorist be able to explain how what appears to be a complex, intentional activity on the part of, say, the interpreter, can be accomplished by some clearly physical means. Contrary to this accusation, Gazzaniga's emphasis has been more on the limited nature of the interpreter than on giving it personhood.

Another more serious objection to this "self-projecting" view of confabulation is that it has a problem with negative confabulations. Why would someone tell a story about his life in which his parents have been replaced by impostors? Gazzaniga's answer is that the interpreter is only as good as the information it receives. This may be correct, but an additional explanation is required as to why frontal verification processes are not able to detect the error.

Is Confabulation Restricted to the Left Hemisphere?

If the left hemisphere is the only one that confabulates, this could be taken as support for the linguistic conception of confabulation; only something that possesses full language ability can confabulate. Several writers have proposed that the right hemisphere is not confabulatory (Ramachandran 1996a; Gazzaniga 2000; Hirstein 2000) and that its function is to provide an accurate picture of reality, partly to balance the confabulatory tendencies of the left hemisphere. The only problem with this plausible hypothesis, however, is that the data refuse to bear it out. Whenever a situation is set up so that the two hemispheres can be assessed for confabulation on an equal basis, the right hemisphere registers as equally confabulatory.

According to Gazzaniga, the right hemisphere is not confabulatory; it "maintains a veridical record of events, leaving the left hemisphere free to elaborate and make inferences about the material presented. In an intact brain, the two systems complement each other, allowing elaborative processing without sacrificing veracity" (2000, 1317). He gives the following piece of experimental evidence to support this claim: "When asked to decide whether a series of stimuli appeared in a study set or not, the right hemisphere is able to identify correctly items that have been seen previously and to reject new items. The left hemisphere, however, tends to falsely recognize new items when they are similar to the presented items, presumably because they fit into the schema it has constructed" (Gazzaniga 2000, 1317; see also Phelps and Gazzaniga 1992; Metcalfe et al. 1995). This schema-construction process has benefits, but its costs become crippling when it is isolated from the right hemisphere: "By going beyond simply observing events to asking why they happened, a brain can cope with such events more effectively should they happen again. In doing so, however, the process of elaborating (story-making) has a deleterious effect on the accuracy of perceptual recognition, as it does with verbal and visual material. Accuracy remains high in the right hemisphere, however, because it does not engage in these interpretive processes" (Gazzaniga 2000, 1317).

Gazzaniga's picture memory task is not adequate to determine whether the right hemisphere is confabulatory, however, because the right hemisphere is *good* at such tasks. Confabulation occurs only when the subject does not have the information or resources needed to answer a question, but answers nevertheless; that is, confabulation is elicited in tasks that the subject is *bad* at.

Confabulation occurs when a perceptual or mnemonic process fails, and the failure is not detected by frontal processes. The proper way to compare the confabulatory tendencies of the two hemispheres is to use tasks at which both of them are fairly bad. When this is done, evidence emerges that both hemispheres are confabulatory. Lu et al. (1997) tested patients with intractable epilepsy, some with anosognosia for hemiplegia and some without, using sodium methohexital to temporarily disable one hemisphere. In half the trials the patients were stimulated on the paretic hand with sandpaper, metal, or rough fabric. Their task was to use the other hand to point to the correct material on a sample sheet, or to a question mark that they had been trained to use to show that they were not stimulated. In the other half of the trials, the patients were not stimulated, so they should have pointed to the question mark. Confabulation was operationally defined as failure to point to the question mark in trials in which no stimulation occurred. This is a nonverbal variant of a procedure that measures confabulation by recording how often the subject answers, "I don't know" (Mercer et al. 1997; see also Dalla Barba 1993). The frequency of confabulation was the same between the two hemispheres, and no difference was seen between patients with anosognosia and those without (Lu et al. 1997).

In an early related experiment, when a split-brain patient was shown a chimeric figure (a figure with half a chair in the left hemifield and half a face in the right hemifield), the patient's left hemisphere would respond verbally by saying that it saw a face, while the right hemisphere would respond by drawing an entire chair (Trevarthen and Sperry 1973). Since neither hemisphere gave any evidence that it was seeing only half of the figure, Feinberg and Roane (1997) interpreted this to mean that the two hemispheres each experience whole stimuli through some sort of completing or interpolating process similar to the one that fills in the visual blind spot. Note that all of the studies showing symmetrical confabulation used a concept in which confabulation is not essentially a linguistic phenomenon. This is necessary, since the right hemisphere is usually unable to express its answer in language. The two hemispheres have a more general confabulatory mode of responding to requests for information, whether or not the answer is in linguistic form.

Is the idea of the veridical right hemisphere mistaken then? It is still possible that the right hemisphere is less confabulatory than the left when it comes to matters more important than simple laboratory tests; the veridical right hemisphere theory might still apply in more significant decisions, perhaps because of their emotional component. It is well confirmed that damage to the right hemisphere can seriously diminish or abolish the SCR (Davidson et al. 1992). In addition, some evidence suggests that damage to the left hemisphere has a sort of releasing effect on the SCR, causing it to increase. If the right hemisphere is the primary initiator of SCRs, this gives it a sort of power, given the relation between emotion and cognition we saw that Damasio and others espouse, in chapter 4. They propose that "somatic markers" play a role in inhibiting certain ill-advised actions (Bechara et al. 2000a). Given that the left hemisphere is the primary controlling hemisphere, production of an SCR by the right hemisphere may be that hemisphere's primary means of affecting the left hemisphere control processes—its primary means of getting the left hemisphere's attention, as it were. This appears in the literature on the autonomic system as the idea that the SCR is the "gateway to consciousness" (Ohman et al. 1993). Since the autonomic system itself is central and not divided into right and left sides, the right hemisphere should still be able to get the left hemisphere's attention in a split-brain patient using the shared autonomic system. As we saw earlier, Gazzaniga proposed and extensively investigated such transhemispheric emotional phenomena.

7.5   Evaluating the Lateral Theories

Are the frontal and lateral theorists referring to the same phenomenon when they speak of confabulation? In the mnemonic conception, according to which confabulation serves to fill gaps in memory, lateral theories are excluded at the outset, since most of the confabulations they are directed at do not involve memory. On the other hand, some frontal or executive theories of confabulation do show a kind of openness that would allow certain lateral phenomena to count as confabulation. There clearly is a difference in what the two patient groups are confabulating about. Memory confabulators make claims about past events in their lives, while split-brain patients confabulate about what they perceive, and about why they perform certain actions and show certain emotions. In this respect, they are more similar to patients with anosognosia and misidentification, whose confabulations originate from a perceptual rather than a mnemonic problem.

Overlap in Frontal and Lateral Phenomena

The anterior communicating artery feeds both frontal parts of the brain that frontal theorists think are involved in confabulation, such as the posterior and medial parts of the orbitofrontal cortex (see chapter 3), as well as the anterior part of the corpus callosum, which lateral theorists claim to be crucial in confabulation, and the anterior commissure. Information is lacking about what percentage of patients with ACoA syndrome sustain damage to the corpus callosum, however. In a report on the neuropsychology of four such patients, three of them had noticeable callosal damage (Beeckmans et al. 1998). Another patient showed signs of cerebral disconnection after ACoA aneurysm (Beukelman et al. 1980). A review of the amnesic syndromes by Parkin and Leng suggested a connection between hemispheric disconnection and the memory problems of some confabulators: "Callosal disconnection may simply be a rare phenomenon in ACoA cases. A more intriguing possibility, however, is that some less dramatic disconnection may be present in other cases, but that appropriate testing to reveal it has not been carried out. Furthermore, such deficits in inter-hemispheric transfer might contribute to the memory impairment in this group" (Parkin and Leng 1993, 125).

After they realized that the corpus callosum does perform valuable functions, epileptologists began removing only those portions serving the affected brain areas. This practice has produced a fairly large set of patients who have had partial callosotomy. If damage to the corpus callosum is important in producing confabulation, this may indicate that its anterior portions are important, because they are the ones fed by the ACoA. Thus patients who have had the posterior parts of their corpus callosum surgically removed should not be confabulatory. It is interesting that both ACoA and split-brain patients sometimes have "alien hand syndrome," a strange disorder in which the patient's hand (typically the left) seems to have "a mind of its own," a kind of disinhibition. Many cases of alien hand involve callosal damage, or infarction of one of the anterior cerebral arteries that are joined by the ACoA. In addition, the HERA model of memory may relate mnemonic confabulation syndromes (Korsakoff's and ACoA aneurysm) to lateral confabulation syndromes. The problem in Korsakoff's amnesia may be that the source information for episodic memories housed in the right hemisphere is unable to causally interact with the left hemisphere centers responsible for communicating that information in natural language.

Emotional Differences between Right and Left

Studies of the emotional reactions of the two hemispheres seem to support the notion of a glib, confabulatory left hemisphere, and a sober, in-touch-with-reality right hemisphere. In a now classic study, Gainotti (1972)

measured mood changes in epileptic patients who underwent a procedure (Wada testing) in which sodium amytal or sodium methohexital was
injected into either the left or right carotid artery to disable the hemisphere
on that side (so that the locus of language functions could be determined
prior to surgery to remove the seizure focus). Patients whose left hemispheres were disabled by the sodium amytal (so that the right hemisphere
was the only one active) were depressed, cried, and were in general de-
spondent. Alternatively, disabling the right hemisphere often produced
joviality, talkativeness, and laughter. The right hemisphere accurately represents the seriousness of the situation—the person has a serious brain disorder and is about to undergo surgery, hence the sadness. Similarly, it may
not be that the isolated left hemisphere was happy, as might appear; it may
simply have been confabulatory and unemotional about the situation.

Awareness in the Left and Right Hemispheres

Recall our noting in chapter 4 that (psychological) skin-conductance responses originate primarily from right hemisphere areas. "Only the right
hemisphere is able to produce an appropriate and selective autonomic response to ... emotional material" (Gainotti 2001, 174). We are typically
not aware of our skin-conductance responses when they happen. "In the
right hemisphere, the generation of the appropriate autonomic response
can be dissociated from the conscious, cognitive evaluation of the elicited
stimulus" (Gainotti 2001, 174). Morris et al. (1998) propose that the right
amygdala responds to stimuli that are masked (hence not consciously perceived by the subject), and the left amygdala responds to unmasked stimuli. This underawareness of the activities of the right hemisphere may help
explain the connection between confabulation and unawareness of illness;
because we are not explicitly aware of right hemisphere activities to begin
with, we tend not to notice when they disappear.

The Left and Right Orbitofrontal Cortices

Disconnection of the left and right orbitofrontal cortices may be an important factor in split-brain confabulation. In this view, the two orbitofrontal cortices have to work together to examine thoughts before they are
made into claims. In an interesting study Cavada et al. (2000) applied a
large stain to the entire left orbitofrontal cortex of a rhesus monkey. They
observed the expected clusters of labeled cells in the superior temporal sulcus and in medial and dorsolateral frontal areas, but the right orbitofrontal
cortex was so thoroughly and uniformly stained it looked as if the injection had been made there. Johnson, who proposed the reality monitoring
theory of memory confabulation described in chapter 3, said that "the fact
that confabulation is associated with damage to medial structures where
there is a relatively high probability that damage will disrupt, if only tem-

porarily, normal interhemispheric communication is consistent with the idea that such self-monitoring ordinarily involves interactions between hemispheres" (Johnson et al. 2000, 378). According to Fischer et al., "the paralimbic frontal structures probably function as bilaterally organized structures and, as such, damage to either hemisphere can lead to similar behavioral and cognitive deficits. Damage to the anterior corpus callosum may be a particularly critical factor ... because of further disruption of bilaterally integrated functional systems" (1995, 27).

The existence of such interhemispheric cooperation is not a new or radical idea. For instance, Johnson and Raye (1998) suggest that the right and left prefrontal cortices cooperate and coordinate their activities during "more reflectively demanding" recall and testing of episodic memories. The brain's overall executive plan runs as follows. The left hemisphere is the primary controlling hemisphere. It enlists the aid of the right hemisphere for various tasks. The right hemisphere contains a large, topographic representation system, including episodic memories and current real-time representations of the body and its surrounding space. A primary task of the right hemisphere is also to simulate other minds. However, these representations are also used for other purposes: finding one's way about, storing important information in memory, and assessing the plausibility of certain events.

In one theory the left hemisphere contains processes that are constantly devising theories and explanations of our present situation (Gazzaniga 2000). In a similar theory, the left hemisphere is capable of "theoretical thought"; it can reason about events that are not the case in the real world (Deglin and Kinsbourne 1996). The right hemisphere is much more concrete. Deglin and Kinsbourne (1996) performed an experiment in which subjects' whose hemispheres were temporarily (30–40 minutes) suppressed by the administration of electroconvulsive therapy were exposed to two types of logical syllogisms. One type began with true premises and a second type began with false premises. When the left hemisphere was suppressed (so that the response required right hemisphere capacities) the subjects did not accept false premises and refused to use them. It is interesting that presentation of the false premises produced strong negative emotions in the subjects: "Noteworthy was a frequently occurring emotional reaction to false premises: 'It's a lie!', 'Nonsense!', 'Rubbish!', or even, 'Doctor, you've gone mad!' This indignation was often followed by corresponding vegetative and motor responses: the subjects flushed, clenched their fists, genuine anger was heard in their voices. (These reactions endorse the intactness of hypothalamic mechanisms after unilateral ECT.)" (Deglin and Kinsbourne 1996, 298–299). These emotional responses indicated that "apparently defense against information which is poor in quality (presumably a right hemisphere function) demands powerful emotional support" (1996, 303). Furthermore, "the right hemisphere seems incapable of the willing suspension of disbelief" (1996, 303).

As a control, under right hemisphere suppression, Deglin and Kinsbourne report that "the same subjects, as a rule, attach no importance to the fact that the information contained in the premises is false and use it with confidence. The subjects' indifference about false premises in this state is striking. Even repeatedly drawing attention to the false information contained in the initial data does not ... prevent them from using it and from arriving at an absurd conclusion, in accordance with the rules of formal logic" (1996, 302).

These findings are suggestive. They present a picture of a left hemisphere that can represent all sorts of contrary-to-fact scenarios, and reason about them without worrying about whether they are true. The right hemisphere, alternatively, is not so much a lover of truth, but planted firmly in what it represents as real. These findings are also consistent with Damasio's work on the emotional components of decision making (section 4.2), and indicate that the emotional inhibition that prevents risky decisions may be primarily the work of the right hemisphere. Recall also my earlier speculation that the SCR produced during lying emanates from the right orbitofrontal cortex (section 4.5). What Deglin and Kinsbourne said about the right hemisphere's intolerance for falsity and a concomitant negative emotional reaction fits the idea that this emotional reaction may be the same one that produces the SCR during lying. It is as if the right hemisphere "hears" the left hemisphere's lie—its deviation from reality— and responds with a powerful negative emotion. The right anterior cingulate may also play an important role in this emotional inhibition process.

7.6   Other Confabulations about Mental States and Intentions

The confabulations that split-brain patients produce belong to a larger family of confabulations about why a subject performed a certain action. When Wilder Penfield electrically stimulated peoples' brains in the 1950s, he was able to cause them to make movements or emit sounds. Sometimes the patients would claim that Penfield was the cause of the movement. They responded with remarks such as "I didn't do that. You did." "I didn't make that sound. You pulled it out of me" (Penfield 1975, 76). In contrast, Hecaen et al. (1949) electrically stimulated the "central nucleus" (this seems to refer to the thalamic mediodorsal nucleus), which caused the patients' fists to clench and unclench or to make "pill rolling" motions. The patients claimed that they had done this intentionally, but were unable to offer a reason for the action.

Delgado's brain stimulation patients also claimed they had performed actions voluntarily, and confabulated a reason why. When Delgado stimulated the rostral part of the internal capsule, this "produced head turning and slow displacement of the body to either side with a well-oriented and apparently normal sequence, as if the patient were looking for something" (1969, 115–116). When the patients were asked why they engaged in those actions, genuine confabulations seemed to result: "The interesting fact was that the patient considered the evoked activity spontaneous and always offered a reasonable explanation for it. When asked 'What are you doing?' the answers were, 'I am looking for my slippers,' 'I heard a noise,' 'I am restless,' and 'I was looking under the bed'" (1969, 115–116). Delgado notes that "It was difficult to ascertain whether the stimulation had evoked a movement which the patient tried to justify, or if an hallucination had been elicited which subsequently induced the patient to move and to explore the surroundings" (1969, 116).

Perhaps, to apply this phenomenon to the findings of chapter 6, when a patient with paralysis and neglect of a left limb is asked to touch the doctor's nose with his left hand and asserts that he did this, after not moving at all, he may be undergoing a sort of hallucinated act of will. He seems to hallucinate not only the execution of the action, but its intentional execution. As Heilman argues (see section 6.2), however, these sorts of hallucinations may not occur in every anosognosic patient. There is also the possibility that at least some of the patients' confabulations are not based on any hallucinations or impressions, but are created only to satisfy the questioner.

The Skeptics
According to some writers, brain stimulation phenomena, together with the evidence from split-brain patients, imply that we often confabulate when asked why we did something or whether we did it intentionally. Wegner says that "the mind is a system that produces *appearances* for its owner.... If the mind can make us 'experience' an airplane, why couldn't it produce an experience of *itself* that leads us to think that it causes its own action? The mind creates this continuous illusion; it really doesn't know what causes its own actions" (2002, 28). Wegner quotes Spinoza who, referring to the ordinary belief of people in free will says, "Their idea of freedom, therefore, is simply their ignorance of any cause for their actions" (1677, part II, 105). Bisiach and Geminiani (1991) speak similarly about two levels present in the human mind, roughly a doing one and a talking one. They posit that this explains the inconsistencies in anosognosic patients, such as the way that many patients deny being paralyzed but never try to get out of bed.

Rolls (1999) claims that confabulation can occur when an action was actually generated at a level lower than that of the one that produced the speech act. But he says that this may happen "sometimes," and does not draw any dire implications for our notion of will or our everyday explanations of action from these phenomena. Rolls says that the following two phenomena are consistent with the idea of multiple routes to action: (1) Split-brain patients may not be aware of actions performed by the non-dominant hemisphere. (2) Patients with damage to the prefrontal cortex can perform certain actions, while commenting verbally that they should not be performing these actions, for example, perseveration on the Wisconsin Card Sorting Test. According to Rolls, "in both these types of patient, confabulation may occur, in that a verbal account of why the action was performed may be given and this may not be related at all to the environmental event which actually triggered the action. It is possible that sometimes in normal humans when actions are initiated as a result of processing in a specialized brain region such as those involved in some types of rewarded behavior, the language system may subsequently elaborate a coherent account of why that action was performed (i.e. confabulate)" (1999, 247).

Rolls then makes an interesting connection to Jackson's idea that the nervous system consists of several levels of perception-action cycles: "This would be consistent with a general view of brain evolution in which, as areas of the cortex evolve, they are laid on top of existing circuitry connecting inputs to outputs, and in which each level in this hierarchy of separate input-output pathways may control behavior according to the specialized function it can perform" (1999, 247).

As opposed to Rolls, who says this might happen *sometimes*, for Gazzaniga et al. these phenomena are pervasive. "Behaviors are being continually exhibited, the origins of which may come from coherent, independent mental subsystems, and these actions have to be and are immediately interpreted by the verbal system. As a consequence, a large part of our sense of conscious reality, we believe, comes from the verbal system attributing cause to behavior" (1977, 1147). It is a happy delusion we are subject to, according to Gazzaniga: "The interpretation of things that we encounter has liberated us from a sense of being determined by our environment; it has created the wonderful sense that our self is in charge of our destiny" (2000, 1320–1321). As I noted earlier though, confabulations that describe events that are negative for the patient (e.g., as in the misidentification syndromes) are a counterexample to this theory.

The Believers
In a view opposing that of the skeptics, sometimes the parts of the brain we are aware of follow, but other times they lead. They lead when we make decisions, but they may follow when we implement them; the conscious decision comes before the work of implementation. There are also several

theorists who posit that the processes embodying conscious states become causally effective when lower-level, more automatic actions fail. Shallice (1988), for instance, posits the existence of something he calls the supervisory activating system that becomes active when routine behavior is not adequate, or a choice needs to be made from several courses of action.

Stuss makes a similar suggestion: "An executive or control function is also hypothesized as depending on frontal brain regions. This executive function provides conscious direction to the posterior-basal and frontal functional systems. The posterior-basal organized integrated fixed functional systems perform efficiently in routine, overlearned situations. In novel, complex, or nonroutine situations, however, conscious direction as provided by the frontal lobes is required. The frontal control system explores, monitors, fixates, selects, shifts, modifies, and judges all nervous system activities" (1991, 68). Timing studies also support the claim that executive areas are active prior to movement. Devinsky et al. for instance say that anterior cingulate activity occurs "well before a movement occurs" (1995, 287).

What then is the difference between intentional, unintentional, and nonintentional actions? Ultimately the definition of intentional action should derive, not from conceptual analysis, but from what we know about the biology of how the brain causes actions. Intentional actions are actions caused in certain ways, while nonintentional actions are caused in other ways, but this may not be a simple dichotomy. If the brain has several different ways to cause actions, it is likely that each way will have interesting implications at the psychological and philosophical levels of inquiry.

These claims about the will's causal efficacy being an illusion may stem from a misunderstanding of the role of the higher-level conscious system in a person's life. The fact that people are sometimes mistaken when they describe actions does not imply that our conscious mental life has little or no causal role in producing our behavior. Much of what the consciousness system does is involved in setting up things so that the automatic processes can work correctly. I decide ahead of time not to give any money to panhandlers, for instance. One reason I do this is that I am not sure what I will do when a panhandler actually approaches me. I consciously program myself to perform the desired action automatically when the time arises. (These objections also apply to the Libet [1996] experiments.)

The problem is that what the skeptics are describing are pathological states, or at least states in which the human brain is being used in suboptimal ways. These conditions, then, are not chronic components of normal human nature. They are human foibles, a misuse of our cognitive equipment, and not the normal case. Another problem is that the skeptics are looking at cases where explanations were created about *past* events. It may be that a better way to understand consciousness is to see its causal role in an action *before* it is committed.

Before I engage in an action, say, going into my boss's office to ask for a raise, I form an intention to ask for a raise. Here is where consciousness plays a causal role. I am aware of forming the intention, rehearsing exactly how I will ask, and what I will say in response to various things he might say. Now, it is quite possible that the idea to ask for a raise was generated outside of my consciousness, and popped into it ready-made. This is different from the actual intention to ask for the raise.

An idea appears in a person's consciousness; then other processes can inhibit that person from acting on the idea. One way we might try out an idea like this is to imagine an event from different perspectives—from my perspective or from a sort of God's-eye view of the event. "How might this look to other people?" we ask. There is no doubt tremendous individual variation in this kind of planning process. I have described it in visual terms, but the same evaluation process can occur with auditory brain areas.

In this view, consciousness functions as a sort of testing ground, where flawed representations are prevented from taking part in the causation and guidance of actions. As we saw in chapter 4, neuroanatomists do not shy away from suggesting roles for both the dorsolateral cortex and the orbitofrontal cortex in a person's conscious awareness, including emotional reactions to stimuli. There are several sources of evidence that the orbitofrontal cortex is an area that can participate in the type of integrated states we saw in chapter 6 posited by some theorists of consciousness. Damasio (1999) suggests that the orbitofrontal cortex contributes certain emotions to conscious states.

The integration of different brain areas into a single conscious state may be achieved through a binding process. A recent class of theories of consciousness identifies it with the binding of disparate brain areas by synchronized electrical oscillations, initiated primarily by the intralaminar nuclei of the thalamus (Llinás 2001; Singer 1999). Anatomists Ongur and Price (2000) affirm that the orbitofrontal cortex has extensive intralaminar connections. It is also worth noting that damage to the intralaminar nuclei can produce neglect (Rizzolatti et al. 2000), and that Mesulam (1981) found that area PG in the rhesus monkey, the equivalent of our inferior parietal cortex, has extensive intralaminar connections. There have also been several claims to the effect that the dorsolateral prefrontal cortex areas responsible for working memory might be vital for consciousness. As we saw in chapter 4, often these theorists include the orbitofrontal cortex as part of this system, responsible for producing certain conscious emotions (e.g., LeDoux 2002).

Obsessive-compulsive disorder itself provides a powerful argument for the efficacy of conscious states. Obsessive states that continuously press themselves on our consciousness are disabling because they put the brain constantly in an inhibit/error-check mode. If conscious states have no real effects, why are the obsessive thoughts so incredibly distracting and debilitating? In their review on the anatomy and physiology of the orbitofrontal cortex, Zald and Kim say that "the inability of OCD patients to inhibit intrusive thoughts and images could reflect a hyperactive working memory process in which internally generated representations (expectancies) are maintained indefinitely in a state of moment-to-moment awareness. In the extreme case, this information might become locked 'on-line,' despite repeated attempts to eliminate the representation" (2001, 59). This indicates a functional role for consciousness. We cannot let certain representations come fully into our consciousness because the critical, inhibiting processes would chop them to pieces. Someone with OCD can be driven nearly insane by a spot on a sheet. This may relate to filling in of the blind spot. Perhaps filling in evolved for similar reasons: so that the checking processes would not be constantly activated by a black spot. Checking processes force conscious attention toward the problem area. Perhaps the idea that the contents of consciousness must be clear and gap free for the checking processes unifies confabulation and filling in.

The literal interpretation of the "self-monitoring" (Stuss and Benson) or "reality monitoring" (Johnson et al.) view we surveyed in chapter 3 may be that the monitoring processes operate on the contents of consciousness. Take whatever representation, x, is part of the current conscious state and create an explanation of x. Test x against your memory. Test x against your conceptual knowledge. Test x against your current representation of your body in space. We do not have to intentionally activate each of these processes. Each one attempts to perform its function on whatever is in our consciousness. When a conflict is found, the anterior cingulate is activated to resolve the conflict (and perhaps also to produce an autonomic alerting reaction). In this view, one function of conscious states is to embody representations that can be examined or monitored by frontally based processes.

Here is a final, evolutionary argument for the efficacy of conscious states: One thing the majority of current theorists of consciousness agree on is that conscious states involve large areas of the cortex and activity among subcortical organs, such as the thalamus and the amygdala. Why would an organ subject to evolutionary forces in the way that the brain has been devote such a huge amount of representational space and computational power to producing a deception for itself? The brain uses a large percentage of the calories we consume. Calories are expensive and often hard to obtain. Why the waste?

7.7 Conclusion

Given what we learned in chapter 4 about how the general function of parietal and temporal areas is to create candidate responses, while the general function of the frontal areas is to check, place in context, and possibly inhibit these responses, it is not surprising that both hemispheres should exhibit confabulatory tendencies when their frontal areas are disabled by disconnection. As we noted in chapter 6, confabulation cannot be explained simply by pointing to the processes in the brain that produce a confabulation and their disconnection from, for example, perceptual processes. There is the additional problem of the "disconnection" of the creative process from the frontally based checking and inhibiting processes that can catch the initial error. Leaving these processes out of the picture can produce the kind of pessimistic view of our knowledge of our mental lives that the skeptics have.

The problem of whether our reports of our intentions and reasons for action might be confabulations is not posed as a problem of memory, but rather as one of knowledge—specifically self-knowledge. The skeptics claim that much or all of the time we do not know what our reasons or intentions are. The skeptics seem to have something epistemic in mind when they call the patients' claims confabulations. Their idea is that the subject believes he has knowledge about the causes of his actions, but in fact has little or none. As Gazzaniga et al. put it: "The verbal system is not always privy to the origins of our actions. It attributes cause to behavior as if it knew, but, in fact, it doesn't" (1977, 1147). The believers tend to be claiming not so much that we know what our intentions, thoughts, and reasons are every minute, but that we *can* know what they are, especially in the case of nonroutine, more cognitively challenging actions. In chapter 8 I argue that, as these disparate views affirm, there is something that unifies the diverse set of confabulating patients: a knowledge deficit.

8 Confabulation and Knowledge

They would be very loath, I fancy, to admit the truth—which is that they are being convicted of pretending to knowledge when they are entirely ignorant.
—Socrates (399 B.C.)

8.1 Confabulation as an Epistemic Phenomenon

I can move my arm.
I can remember what happened to me yesterday.
I can recognize my family and friends.
I know whether I did something on purpose or not.
I can see.

How could anyone be wrong about any of these claims? They are so basic to our belief systems that they sound odd when explicitly stated. Each of these is a claim to a specific sort of knowledge, and each claim originates within a certain knowledge domain. The brain's implementation of each knowledge domain is subject to characteristic confabulation syndromes:

1. Knowledge of the body and its surroundings (anosognosia for hemiplegia and/or hemianopia—loss of half of the visual field)
2. Knowledge of recent events involving oneself (Korsakoff's syndrome; aneurysm of the anterior communicating artery)
3. Knowledge of other people (the misidentification syndromes)
4. Knowledge of our minds (split-brain syndrome, brain stimulation cases)
5. Knowledge derived from visual perception (Anton's syndrome)

There is something all cases of confabulation have in common: a knowledge deficit. The failings in each type of confabulation are epistemic—the confabulator is expressing an ill-grounded thought that he or she does not know is ill-grounded. Epistemology is a branch of philosophy whose practitioners study knowledge itself; its name comes from the Greek word "*episteme*," to know.

What constitutes your knowledge? Part of it is embodied in your memory, both in the semantic and episodic-autobiographical domains. Another part of your knowledge is in your current perceptions—of your surroundings, but also of your body, and of your mind itself. Many writers on confabulation accept that there are two disparate sets of confabulation syndromes, one involving memory deficits, and another that is "likely to involve specific deficits in perceptual processing (Capgras' syndrome, reduplicative paramnesia, Anton's syndrome and confabulation associated with hemiplegia)" (Johnson et al. 2000, 263). But there need not be a dichotomy between confabulation about memories and confabulation about perceptions; both memory and perception are knowledge domains.

Confabulatory patients have an epistemic problem because their brains do not produce claims with probabilities as high as those it would produce if they were not compromised. When asked a question, they toss out ill-grounded answers. Thus we should read the remarks about the fatuousness and glibness of confabulators partly as epistemic remarks. Confabulators are not worried about something they should be worried about—the truth of what they are saying. As we saw in the preceding chapters, confabulations are the product of two errors, both of which are epistemic. First, an error occurs in some knowledge process that causes an ill-grounded thought (or more broadly, an ill-grounded representation) in that domain. Second, frontally based processes that should function to verify that thought also fail. The failure of this second process is the phenomenon described by some writers as a failure to be self-reflective, to self-monitor, or to self-correct (Stuss and Benson 1986; Johnson et al. 2000).

The brain employs many checking processes:

- Checking against one's autobiographical memory
- Checking for internal contradictions (one type of this might be done using semantic memory)
- Checking against one's factual knowledge (i.e., checking against semantic memory)
- Checking against one's body representations
- Modeling the current thought's effect on other people
- Checking by observing one's audience
- Checking the veridicality of one's current visual images
- Assessing the reward value of the current thought
- Discerning logical and causal consequences of the current thought, and assessing their reward value
- Determining one's level of confidence in the current thought
- Developing intentions to act based on the current thought

Apparently some of these checking processes develop naturally, such as the checks against memory, while others are explicitly learned. Assessing the logical consequences of a thought, for instance, needs to be learned in the case of certain inferences and fallacies. The primary function these checking processes share is epistemic. They ensure that the thoughts that are selected as beliefs have the highest truth probabilities the person's cognitive system can achieve. Each process has its domain, a circumscribed set of representations it can usefully check, but there is a large degree of overlap in these domains. What all cases of confabulation share is a failure in some epistemic domain, coupled with this checking failure.

The theories developed by philosophers for examining knowledge claims and criteria—the contemporary field of epistemology—will thus be of use to us in fleshing out an epistemic approach to confabulation. One of the products of debates among contemporary theorists of knowledge in philosophy has been a detailed analysis of what constitutes a proper warrant, or ground, for a knowledge claim. Some of these theories will prove useful in explaining exactly what it means to call a belief or thought ill-grounded.

Alvin Goldman's landmark book *Epistemology and Cognition* (1986) was an important step in the radical (for philosophy) project first sketched out by W. V. O. Quine (1969), of "naturalizing" epistemology—constructing a scientific epistemology. One objection to this project from more traditional philosophers has always been that epistemology is about how we *should* obtain knowledge and make knowledge claims, while science can only tell us how we *do* obtain knowledge and make knowledge claims. Goldman made the important point, however (which is rather obvious when it is explicitly stated), that how we do obtain knowledge is highly relevant to the question of how we should obtain it. In normal circumstances we cannot exceed the bounds of the knowing apparatus given to us by nature. What we would be wise to do is try to understand how best to use the apparatus we have. A study of the ways in which this apparatus breaks down yields valuable insights into what constitutes a well-grounded as opposed to an ill-grounded belief.

8.2 The Neuroscience of Confabulation

At this point we have enough information to begin to sketch the structure of the systems that are disrupted in confabulation. Despite the many different lesion areas and theories of their function surveyed in chapters 3 through 7, a coherent picture emerges. Damage to areas involved in the implementation of several knowledge domains, coupled with damage to orbitofrontally based monitoring processes, can cause confabulation.

Confabulation in response to a question begins with the question itself. The question has to tap a knowledge domain compromised in the patient. Then either the processes associated with that knowledge domain produce a faulty representation, or other, nonstandard, processes produce a faulty representation. Why, though, does the left hemisphere insist on filling that gap when no correct memory surfaces? Apparently our normal state, in which we are able to say that we remember nothing if we do not have a correct answer, is not due merely to nothing coming into consciousness, but to forces actively suppressing incorrect thoughts from becoming conscious. Schnider's (2001) hypothesis about the brain actively suppressing irrelevant memories (see section 3.5) is also consistent with this idea. Another way to know that you don't know the answer to a question is by being able to reject every answer your mind comes up with. Once a faulty representation has been created, there is still a way that confabulation can be prevented: The problem with the representation can be detected by one of several checking procedures.

The data from anatomy, physiology, and lesion studies indicate that many of these checking processes are carried out by neural circuits running through the orbitofrontal cortex (including orbitomedial areas). Some of the functions performed by these circuits, such as checking against one's body representations and observing one's audience, appear to involve mainly right hemisphere processes. Bifrontal processes in which the left and right orbitofrontal cortices work together to evaluate a given thought may also be crucial to this checking phase. The failure to pass a check can be registered through processes that involve the orbitomedial frontal cortex and the hypothalamus or other autonomic centers. Several of the areas capable of initiating autonomic responses have also been implicated in confabulation, including the orbitofrontal cortex and the right inferior parietal cortex. These emotional-autonomic connections are crucial if the failure to pass a check is to stop a contemplated action from forming into an intention to act. Merely knowing that a representation is faulty is apparently not enough to keep some people from making a claim based on it. The data show that in the human brain, stopping a candidate action from becoming an intended, then executed, action often involves the emotional systems.

There is a common finding in confabulators of a hyporesponsive, or unresponsive autonomic system. Korsakoff's patients show damage to noradrenergic structures (section 3.3). People with orbitofrontal damage (section 4.5) and sociopaths (section 4.4) show reduced autonomic activity to certain types of stimuli. Capgras' patients show reduced skin conductance responses to the sight of familiar people (section 5.3). Patients with neglect and flattened affect show either greatly reduced skin-conductance activity (section 6.3) or none. A system of connected areas spanning the medial orbitofrontal cortex and the hypothalamus seems to be critically involved in an emotion-based inhibiting system. Confabulators have lost part of the emotional component in thought. They do not experience the feeling of doubt, or other negative emotions that the brain attaches to thoughts to signify that they have failed a check. Negative emotional tags cannot be attached to a thought, so it becomes a belief and sometimes also leads to an intention to act. Damasio's orbitofrontal patient E.V.R. (see section 4.2) is important because what happened to him shows that the brain's executive processes become powerless without the proper emotional force behind them.

What we know about the orbitofrontal cortex has to do with its sensitivity to reward and punishment (e.g., Rolls 1999). The argument that the orbitofrontal cortex functions as a reliability-improving process depends on a basic conversion, from the negative and positive reward values this cortex specializes in, to truth and falsity. There are some obvious connections; finding that a belief of yours is false is often an unpleasant experience. The autonomic response that stops certain actions also indicates that the current thought has a low value. False thoughts are one type of low-value thought, but ultimately it may be difficult for us to distinguish between "this belief is false," and "this belief led to negative results." Conversely, we tend to retain beliefs that lead to reward, even when we suspect they are false.

While confabulation involves two types of malfunction, this does not imply that two separate events in which the brain is damaged are required for confabulation to occur, since there may be times when a single event damages both processes, such as stroke damage in the right inferior parietal cortex, or an aneurysm of the anterior communicating artery. Representation-creating and representation-checking processes can share a malfunctioning process, compromising both, which I suggested may be the case with the inferior parietal cortex. The brain has a large network of interconnected structures, and sites of damage relevant to confabulation are part of this network. The orbitofrontal cortex, the areas around the superior temporal sulcus, and the inferior parietal cortex are some of the areas affected.

Damage to other areas may cause confabulation by disrupting processing in these primary areas. Severing the corpus callosum may prevent the orbitofrontal cortices from intercommunicating in the way they need to in order to fulfill their checking functions. Similarly, damage to the thalamic mediodorsal nucleus may disrupt processing in the orbitofrontal cortex, with which it is heavily interconnected. Damage to the caudate nucleus may prevent the orbitofrontal cortex from acting to stop behavior, while damage to the hypothalamus may interrupt the autonomic and emotional processes needed to effect an inhibition.

8.3   Creation and Checking of Mental Representations

The brain is filled with representations. Some of them are imagelike and are contained in the brain's many topographic maps. Others are more conceptlike, such as those that comprise our semantic memory. I will refer to the first type as analog representations, and the second type as conceptual representations. The representations that we are aware of come from two fundamental sources: perception and memory. The brain also has processes that can take these basic representations and combine them, compare them, check them, and even claim them, i.e., the processes that we know as thinking, deciding, planning, asserting, and so on.

Confabulation arises from a creative act that produces the claim itself. In the confabulator, these creative processes produce claims with unacceptably low truth probabilities. Next, something causes the brain to miss its second chance to get it right—it fails to detect the defective claim. The way to detect a defective representation is to check it. Checking itself can be implemented fairly simply. If you can combine a representation-evaluating process with a behavior-inhibiting process, the combination can function as a checking process. Checking often proceeds as follows. An object is perceived and representations of it are thoroughly processed. This would involve activity along the entire ventral cortical moiety, from the occipital cortex, through the temporal lobes and amygdala, to the orbitofrontal and medial frontal areas. Limbic responses are generated, based on motives and on past experience with that object. Additional representations can also be generated as part of the evaluation process, including models of what might happen in the future, and the reward values associated with those representations are weighed. In the course of this processing, if a representation is generated that has a high negative value, the cognitive system normally initiates an inhibitory process that prevents any action that might bring about those negatively weighted consequences.

When I judge something to be false or implausible, often what happens is something like this: I consider the thought, that is, I hold it in consciousness, and then, rather miraculously, if any other representation in my brain conflicts with it, it speaks up. This process proceeds as follows. First, create a thought; second, check the value of the thought. If the value is high, continue the thought by using it to generate another thought. If the value is low, stop the thought from producing intentions. "Checking the value of a thought" is shorthand for "checking the value, to the thinker, of the state of affairs the thought represents." The thoughts in our minds can come from memory, or from perception, or they can be generated from other thoughts in many different ways. To each thought is attached an emotional valuation.

Creativity and doubt keep each other in delicate check in the healthy, rational mind, in this picture. As we saw in chapter 2, nature's design system favors pitting countervailing forces against one another. In the autonomic system, for instance, the parasympathetic and sympathetic systems balance each other. From an engineering point of view, this is a good design because it allows the system to dynamically adjust itself in either direction. What the phenomenon of confabulation shows is that the brain's checking ability can be damaged while its creative ability remains intact. This suggests that there are two basic types of malfunction in the entire system, a malfunction in this checking ability and a malfunction in the creative faculties themselves. This implies there are three abnormal types: (1) In the first, generation is intact but the checking process malfunctions. Confabulation-like behavior in normal people and other disinhibition syndromes fall into this group; Tourette's syndrome (involuntary motor and vocal tics) might also fit here. The thoughts generated are often normal enough, but normal people do not express them. (2) In the second type, the generation process malfunctions but checking is intact. Broca's aphasia may fit this description; akinetic mutism might be a failure to generate thoughts. In obsessive-compulsive disorder, generation is malfunctioning, in that no new thoughts are being produced; the checking processes are intact and hyperfunctioning. (3) Finally, in the third type both generation and checking malfunction. A global malfunction in both processes might result in severe, or "spontaneous" confabulation, or in schizophrenia. If the failure at either level is local, there may be domain-specific varieties of confabulation in which generation malfunctions, e.g., the claims of movement by anosognosics. Confabulation can be domain specific or it can be broader, encompassing topic areas relating to the patient, or it can be even broader to include both personal and impersonal knowledge domains.

Creation Phase
Apparently two people can sustain the same type of damage (e.g., an aneurysm of the anterior communicating artery) and one will confabulate and the other will not. The nonconfabulator lacks the ability to create the thoughts which when expressed are confabulations. This positive factor, coupled with the two epistemic errors, means that whether a given person confabulates is a function of three independent factors: (early) damage to the epistemic system, damage to the checking system, and positive claim-creating tendencies. Malfunction in the generation of claims is subject to more or less severe errors. Plausible but false claims may be created. With more severe damage to the generation systems, however, implausible claims may be created. If the person is still going to be confabulatory, though, the systems for generating thoughts cannot malfunction so severely that they produce no thoughts at all.

What I call the creative stream has been referred to as "the language axis" (Joseph 1986), "the interpreter" (Gazzaniga 1995a), "the language system" (Rolls 1999), or simply "language" (Bisiach and Geminiani 1991). For confabulation to occur, the creative faculty must compose a claim that will elude the existing checking processes, before the claim can be publicly expressed. The representations generated can be neutral, positive, or negative, in the person's way of assigning value to them.

Paranoia involves the production of negative representations. As with confabulation in general, we might distinguish between plausible and implausible paranoia. Someone who believes that her husband is cheating on her, even without any real evidence, has plausible paranoia, while someone who believes that her husband is the leader of an alien invasion of Earth has implausible paranoia. The creative stream can be characterized in several ways, all of which are present in confabulation: wishful ("My health is fine"); paranoid ("That impostor is following me"); irrational ("It is World War II and it is 1972"); and egocentric ("That famous conductor is in love with me").

Recall from chapter 2 Hughlings Jackson's enduring idea that lower levels of function might be released by damage to higher levels. Speaking about a group of Korsakoff's patients, Talland says, "A marked degree of intraindividual consistency in confabulatory behavior or its total absence, supports the view that this symptom is an interaction effect of the amnesic syndrome and basic personality structure" (1961, 380). A difference in their creative streams explains the difference between disinhibited autistic people and sociopaths, for instance. Both are bad at mind reading and both are disinhibited, but the creative stream of the sociopath generates selfish, malevolent thoughts, whereas the creative stream in autistic people does not generate such thoughts. This creative element plays a large role in what we saw earlier investigators refer to as the contribution of the patient's premorbid personality. The idea of releasing implies that the people in whom the inhibitory processes have been most active are the ones who will change the most after damage to the checking and inhibiting processes, such as lobotomy. In the absence of these processes, wishful thinking can proceed unhindered. This explains why many confabulations express desirable states of affairs for the patient; human thought in general is naturally selfish and wishful. These tendencies need to be curbed in order to increase the effectiveness of our thought in a world which is, alas, completely unsympathetic toward our desires.

Checking Phase

The creation of adequate, useful representations by the human brain is not an all-or-nothing affair, but exists in degrees. Farah (1994), for instance, speaks of "degraded" representations. Recall also from chapter 7 what Geschwind said about how there need not be a full disconnection between hemispheres; a bad connection is sufficient to bring about the lack of information that in his view causes confabulation: The left hemisphere "may respond to incomplete information coming over subcortical pathways which are inadequately extensive to carry all the information about the seen object; this may lead to errors which are less bizarre than those in which no information at all gets to the speech cortex" (Geschwind 1965, 599).

One interpretation of some of the phenomena we have seen is that degraded representations play a role in producing confabulation as follows. A representation can only be "filtered" by frontal processes if it is complete and clear enough for the checking processes to operate on it. This connection, between a representation problem and confabulation, is so strong we build it into our conceptions of anosognosia and denial; we nonchalantly make it when we say "neglect, with denial." The checking processes are a second line of defense against committing dangerous, offensive, or ineffective actions, including making ill-grounded and false claims. But perhaps these processes can only work on representations of proper quality. Beliefs based on degraded representations are ill-grounded. We can capture the hypothesis indicated here in the *degraded representation principle*: If a person has a diminished capacity to represent events of type x, that person will be more likely to confabulate about events of type x.

What does it mean for one brain area to check a representation embodied in another area? Neuroscientists employ the concept of one brain area monitoring the activity of another area. This is not meant to imply, nor need it imply, that some personlike entity, or homunculus, is observing a certain representation. At a minimum, all it means if one says that area A monitors a representation in area B is that area A stands in a certain type of functional relationship to area B. It typically also means that area A is in (near) constant causal contact with area B. It also usually implies that area A initiates another process when area B contains a certain representation.

But, one might ask, do the checking processes interact with the representations *as* representations—are they representations *for* the checking processes? This brings up a problem about the nature of the relation between checking processes and what they check. Notice, however, that merely saying that a checking process acts on a representation does not commit one to the claim that it is a representation *to* or *for* the checking process. The computer systems used by blind people that can scan a book and "read" its contents out loud are performing certain functions on representations, but they are not representations to or for the reading machines. The thoughts and beliefs that are checked are representations for the agent; one need not also claim that they are the thoughts and beliefs of the checking processes.

The anatomists seem satisfied that the structures and connections needed for such monitoring are in place. We have been focusing on visual representations and conceptual representations, but auditory representations are also monitored, which is merely to say that in different ways, we talk to ourselves. The anatomist Barbas described some of the brain areas involved in this process: "One example of the intricate use of sensory information by prefrontal limbic areas can be seen for caudal medial areas, which are interconnected with auditory cortices.... The circuits linking medial prefrontal and auditory cortices appear to have a role in monitoring inner speech and in distinguishing external auditory stimuli from internal auditory representations through distinct patterns of activation. Moreover, these specific functional relationships are altered in schizophrenic patients who experience auditory hallucinations" (2000, 326). Pandya and Yeterian offer a similar description that applies to all modalities: "In order to allow for the monitoring of incoming stimuli, a system of feedback pathways must be in place." These pathways include "long association connections that project back to post-Rolandic regions from which the frontal lobe receives input" (2001, 50).

Inhibition

Perceiving a desired object in one's personal space often gives rise to the thought of reaching for it. The checking processes must be capable of stopping this behavior in many situations. The brain seems to possess both "acute" and "chronic" inhibitory processes. It can inhibit an action just once at a certain moment, or it can always inhibit that action. Many confabulatory patients seem to be able to stop confabulation when expressly instructed to, at least for a moment. They may briefly admit they are paralyzed, or that they have no idea why their left arms did what they did, but they then uniformly lapse into their confabulations moments later. The patients seem to lack the ability to sustain the correction. Perhaps certain thoughts need to be permanently assigned a negative tag because when they were allowed to lead to actions, undesirable results ensued. This might mean that the claim that the thought gave rise to is false, inappropriate, or merely unpopular.

The Capgras' patient of Mattioli et al. showed the kind of bland admission one can sometimes wring from a confabulating patient, and how quickly it disappears from their minds: "In a general way, when confronted with reality, he would reluctantly admit it for a moment without any sign of surprise or insight, but soon would go back to his former interpretation" (1999, 417). Gassell and Williams (1963) tested patients with hemianopia as follows. The examiner sat facing the patients, and with their eyes fixated on his nose, asked the patients whether they saw his entire face. As the subjects were fixating, any view they might have had of the examiner's face in the suspect hemifield was occluded by a screen. Thus tested, twenty-eight of thirty-five patients insisted they saw the examiner's entire face—a confabulation. However, Gasell and Williams found that if they merely urged their patients to adopt a more "analytical attitude," the patients were less likely to report complete stimuli. Perhaps in those cases in which the person briefly admits confabulation the *doctor* is acting as the person's orbitofrontal cortex normally would. These types of phenomena also support the idea that confabulation involves damage to a large processing system that can temporarily readjust itself in different ways. When this process successfully readjusts itself in the long term, the patient stops confabulating.

8.4   Defining Confabulation

Philosophers of language distinguish two types of definition: a logically airtight kind, known as a definition by necessary and sufficient conditions, and a looser type one can fall back on if it is impossible to construct the tighter one. A definition by necessary and sufficient conditions usually consists of a list of criteria that are individually necessary and jointly sufficient to characterize the concept in question. A bachelor, for instance, is (1) unmarried, (2) adult, and (3) male. Each of these conditions is necessary, and together they are sufficient to make someone a bachelor. Testing of candidate definitions becomes a matter of finding something that fits the concept but not the definition, or vice versa. The looser type of definition is sometimes called a family resemblance definition, after Wittgenstein, who compared it to a list of characteristics of members of a certain family. There might be a characteristic nose, for instance. Not every family member will have that type of nose, but there will be other characteristics that the person will have, such as a type of eyes, ears, forehead, or chin. Thus, no particular characteristic is necessary, but there are several important criteria. There is a family of these interrelated criteria, and things falling within the concept possess a threshold number of these to a certain degree.

We can begin with an attempt at a necessary and sufficient definition, and see whether we can create something that demarcates the right territory, based on what we have learned from chapters 3 through 7 about the facts of confabulation. Even if later we find phenomena we want to call confabulation that do not fit the set, there is still room for the feature in question to participate in the definition, just not as a full necessary condition:

Jan confabulates if and only if:
1. Jan claims that p. (e.g., Jan claims that her left arm is fine.)
2. Jan believes that p.
3. Jan's thought that p is ill-grounded.
4. Jan does not know that her thought is ill-grounded.
5. Jan should know that her thought is ill-grounded.
6. Jan is confident that p.

The next section discusses the criteria in greater detail.

Candidates for Criteria

1. Jan claims that *p*   Why not just say that Jan *asserts* that *p*? Primarily because "assertion" has a strong connotation that speaking is involved. The more general notion of claiming covers speaking, copying drawings, pointing at pictures, selecting shapes, etc. Recall from chapter 6 that Geschwind (1965) coined a phrase, "confabulatory completion," to describe additions that patients with a visual field defect made to drawings. The absence of this claiming feature seems in some cases sufficient to raise questions about whether the phenomenon in question should really be considered a confabulation. Feinberg and Roane, for instance, speak of "confabulatory completion of the blind spot" (1997, 74) in normal vision without noticing that an important element of confabulation is gone: the making of some sort of claim by the person.

2. Jan believes that *p*   We have an implicit understanding that if a person claims or asserts something, he or she believes it, unless he or she is lying, joking, being sarcastic, etc. Violation of this maxim results in something called Moore's paradox, from the philosopher G. E. Moore: Is it coherent to say "*p*, but I do not believe that *p*"? Surely not; it is bizarre to say, "Today is Tuesday, but I do not believe that today is Tuesday." Most of the clinical evidence is in favor of the idea that confabulators are sincere in their claims. Acting on the basis of a thought is powerful evidence that the person accords that thought the status of belief. The frequency of cases cited in which confabulators act on their claims supports the idea that their claims report real beliefs. Another point in favor of a belief criterion is that if a clinical confabulator is probed further, he will affirm that his claim is a belief of his.

I have noted that confabulators are often quite rational in their unaffected domains of knowledge. They can have meaningful conversations, and most of their beliefs and perceptions are correct. The philosopher Donald Davidson questions the idea that a person could have an irrational cluster of beliefs while being sane and rational in general. "If you think, as I do," he says "that the mere possession of propositional attitudes [i.e., beliefs and desires] implies a large degree of consistency, and that the identification of beliefs depends in part on their logical relation to other beliefs, then inconsistencies impose a strain on the attribution and explanation of beliefs" (Davidson 1998, 5). But how else can we explain why the patient with Capgras' syndrome won't sleep with his wife anymore? He won't sleep with her because he believes she is someone else. "Why don't these patients report the impostors to the police?" a skeptic might ask. Some do (Christodoulou, 1977).

We should let data inform theory here and not vice versa. People really do have irrationality in restricted, often odd ways, such as Anton's

(1899) original patient who hallucinated only in his neglected hemifield. There is even a subclass of delusions called content-specific delusions, which center on a specific content area, including Capgras' and reduplicative paramnesia, but also delusions of infestation (Richardson and Malloy 2001). The idea of a homogeneous web of belief is natural. Nevertheless, there are divisions and partitions in our belief systems, and in our brains' systems for checking beliefs, that allow local irrationalities to remain local.

There is also evidence against the claim that confabulators are expressing real, enduring beliefs, however. Behavior that is inconsistent with what a person says is troubling. Sometimes such behavior can be dismissed as being due to weakness of the will, hypocrisy, or self-deception. Recall from chapter 6 that verbal denials of illness and attempted behavior inconsistent with that illness are apparently doubly dissociable. Bisiach and Geminiani point out that patients who verbally deny being paralyzed do not object to being confined to bed. "In contrast," they say, "patients who verbally admit of paralysis on one side may attempt to stand and walk" (1991, 19). Even in cases of such inconsistency, however, we can discern what the person believes. When the person with obsessive-compulsive disorder, for instance, goes back to check the lock on her front door again, there is little problem in saying that she believes (even knows) that the door is locked. The problem is that she cannot stop herself from executing the checking action. If asked about her confidence in her claims made on any particular occasion, a confabulator will almost always express high confidence. If the rationality of such patients is largely intact, this is an argument for taking them at their word when they avow and reaffirm their confabulations. Part of what it is to be a rational person is that one knows what one does and does not believe.

Another objection to making the belief criterion necessary is the way that confabulators may give different (sometimes inconsistent) responses to the same question asked at different times. The states that give rise to their claims do not seem to have the permanence of normal beliefs. They are perhaps more momentary whims or fancies than genuine beliefs. As is often the case with the mentally ill, folk psychology is strained to the breaking point as we attempt to describe them. Together, however, these problems indicate that the belief criterion may not be a necessary one. Even if this is true, belief may still be important for defining the concept of confabulation.

We also assume here that Jan expresses her belief by claiming that p. There are rare times when this can fail; for instance, someone could say something in a foreign language that corresponded to something he believed, yet not be expressing his belief in saying that. One way to respond to this is to say that the belief that p had the normal causal role in the person's act of claiming that p. The belief normally works together with the motive and intention to answer the question to produce the verbal act of confabulation. Normally we regulate this process carefully, but in confabulators, the motive seems to enlist the current contents of consciousness, whatever they are, and begin the process of making a claim. Note though, that the "clutch" normal persons keep disengaged between this motive and the act of claiming has to do with belief. We ascertain whether the current thought is a belief we are willing to assert and act upon. If the damaged processes that allow motives to run so easily to verbal expression in confabulators are belief-manipulating processes, this would again tend to support a belief criterion.

3. Jan's thought that p is ill-grounded   A malfunction in a knowledge-gathering process, either in perception or memory, causes ill-grounded representations to appear in the confabulator's mind. Typically these are mistakenly accorded belief status. As beliefs, their causal history is abnormal and/or suspect. Each type of confabulation syndrome will have a different explanation for why the thoughts in the crucial domain are ill-grounded. In Anton's syndrome, the problem is blindness, in Korsakoff's syndrome, the problem is typically the loss of source information for memories. Anosognosia seems to involve deficits in representing the body, and Capgras' and the other misidentification syndromes involve mind-reading deficits, if the approach taken in chapter 5 is on target.

One way that a representation in the brain can be ill-grounded is by being produced by a process poorly suited to the domain involved. Much of the time the thoughts that give rise to confabulations are not generated by processes that would normally generate a thought with that content. When an anosognosic patient claims that his left arm is strong, the belief he expresses does not have the right causal history. If I ask you whether your left arm is weak, you will look down at it, flex it, and sensing nothing wrong, reply that it is just fine. Your brief check involves, I suggest, important activity in some of the brain's many somatotopic maps and the processes and structures that support them. When you answer that your arm is fine, this answer is the result of a causal process that in the early stages involved somatosensory and somatomotor maps, including maps situated in the right inferior parietal cortex. At some later stage the information contained in these maps is put into concepts capable of being expressed by the sentence, "Yes I can move it."

Another way that a thought that p can be ill-grounded is when the thought that not p played a certain type of role in its genesis. Being approached by a doctor as you lie in bed and asked whether you can move your left arm is sufficient to produce the unpleasant thought that you might be paralyzed, at least for a normal person. We all seem to have a natural reaction to form a belief contrary to a thought that is repellant to us. Confabulators may not show the autonomic signs of fear, but they may still experience fear of paralysis at a more intellectual or cognitive level, and this itself may generate the thought that they are fine.

4. Jan does not know that her thought is ill-grounded Ask a confabulator how he knows what he claims and you are likely to get another confabulation. This criterion is intended to capture the failure of the checking processes because they are the means by which normal people know that their thoughts are ill-grounded. If I know that my memory for recent episodes is nonexistent, for instance, I know that any putative memories that pop into my mind are ill-grounded. But, as Barbizet puts it in the case of memory confabulators, the problem is that "the patient cannot remember that he cannot remember" (1963, 127). Medial temporal lobe memory patients do not confabulate. They do not know what they did yesterday, but they do not confabulate because they know that they do not know what they did yesterday. The researchers who devised "Don't Know" tests of confabulation (Mercer et al. 1977; Dalla Barba et al. 1990) were assuming something similar to this criterion. To know that your thought is ill-grounded is to be open to giving an "I don't know" reply.

In his book *The Neurology of Thinking*, D. Frank Benson says, "in all three conditions—reduplicative paramnesia, Capgras' syndrome, and spontaneous confabulation—decreased ability to monitor the obvious incorrectness of a belief and/or to act on corrected information represents the key dysfunction" (1994, 220). We have ways of knowing whether the thoughts that appear in our minds are well-grounded, including the checking processes described earlier. Some of these are specific to the particular knowledge domain involved. In the perceptual domain's visual modality, we know that putative visual images that are degraded or fleeting are suspect. In the memory domain, we know that if we always mis-remember names, there is reason to suspect our beliefs about peoples' names. Other ways of discerning ill-groundedness can be applied across all domains, such as basic notions of consistency. Noting that what one claimed just now is inconsistent with something one said two minutes ago is another way to distrust one of the two claims. This type of failure may be due to a basic memory problem, in that the patient has simply forgotten the earlier claim, but these are memory processes that are vital for us in exercising our optimal epistemic capability.

Thoughts are powerless unless they are affirmed by emotions. The knowledge that a thought is ill-grounded needs to enlist other, emotional processes to stop a person from making claims or executing other actions based on it. If I learn that my belief is ill-grounded, does that automatically make me change the belief? Not always. Frontal patients often know they are answering incorrectly but cannot stop, as the phenomenon of perseveration attests. Damasio's work seems to confirm the idea that some emotional component is needed before the belief is actually dislodged and any actions based on it inhibited. The possibility of mind-reading impairments in confabulating patients is relevant here. The patients have a powerful reason to question the groundedness of their thoughts, which comes from the people they interact with. When an anosognosic patient tells a therapist he doesn't need any help because he is fine, he gets firm disagreement; when he tells his family he'll be mowing the lawn later today, he is met by expressions of shock. Anosognosics have no reluctance to contradict their doctors about their condition. They either do not perceive these audience reactions, or they do not attach any significance to them.

One objection to including this criterion is that it is redundant, given the second criterion, Jan believes that *p*. If you believe a thought, by definition, you consider that thought to be well-grounded. Given the possibility that the belief criterion is not necessary, however, anything implied by that criterion needs to be supported independently. There are also those cases where people knowingly hold ill-grounded beliefs. Confabulators often seem to go through a phase before they fully recover where they form a belief, then remember that it is ill-grounded. When asked what he did yesterday, a recovering Korsakoff's patient might answer, "Well it seems to me that I was at a baseball game, but I've got this memory problem, and I've been in the hospital for a while now, so it probably didn't happen." There are also conceivable normal cases where people knowingly hold ill-grounded beliefs. Someone might believe that a certain lottery number he has just bet on will win because he saw it written on an overpass just as he was considering what number to bet on. He knows that the provenance of his belief is rather suspect, but he still clings to it. Some religious people would affirm that they both believe in God and know that this belief is ill-grounded; that is the function of faith, they might argue.

5. Jan should know that her thought is ill-grounded   The use of "should" here involves no more than the claim that an optimally functioning car should move down the road. In an optimally operating human mind, the confabulator's thought will be detected and rejected, and not allowed to initiate actions. The person who owns this optimally operating mind would know that the thought that *p* is ill-grounded, and not grant it belief status. Normal people who are blind know that any visionlike images that appear in their heads are mere images. But, how difficult must it be for a person to discover that his belief is ill-grounded? With the sort of beliefs held by confabulators, it would be quite easy for a normal person to make this discovery.

A patient should know that he or she has sustained brain damage, and that the damage affects the groundedness of certain thoughts. Both pieces of knowledge are important because a patient might accept that he or she has a disability, but not that it causes thoughts to be ill-grounded. He could even know that his memory is bad, but still be confabulating because he repeatedly claims that his *present* memory is real. In confabulation, the mind is working suboptimally; processes that evolved to have quite definite functions are not operating. To say that the confabulator should know that he does not know is to say that if his mind were working properly, he would know. One reason the confabulator should know that his belief is ill-grounded is the reactions of other people to his claim. Split-brain patients, for instance, should know that their doctors know they don't know why their left arms are doing what they are.

We need to be careful not to define away a genuine problem here—the one raised in chapter 7 about how statements of intentions and actions by normal people might be confabulations. To claim that people are confabulating is to put them at an epistemic fault, something that is in their power to correct. There may be a percentage of the population whose introspective reports are confabulatory, but this does not mean that all such reports are. Given the ability to detect mental events in the brain, at some point in the future, it might he possible to determine the percentage who confabulate and compare their mental events with their statements. This can involve examining the nature of the causal relations between the mental event itself and the mental events that lead to the utterance. These two events need to be properly related for a statement to be accurate.

We have a knowledge deficit when it comes to the visual blind spot: I don't know what is in that part of my visual field, and (usually) I don't know that I don't know that. I suggested earlier that the brain's act of filling in the visual blind spot (Ramachandran and Churchland 1994) alone does not count as confabulation, since the person has not made any claim. However, if I ask you what is in an area that corresponds to your blind spot, and you answer, "Just more white wall," are you confabulating? It depends on whether we think that you should know you have a blind spot there. Normally, people do not know this, and it seems unreasonable to find them at fault for not knowing.

6. Jan is confident that *p* Confidence is both an emotional and an epistemic concept. One way to bring out the epistemic dimension is to note that if you ask a confabulator whether he *knows* what he is claiming, the answer will uniformly be "yes." The emotional side of confidence relates to the finding of low autonomic responses in confabulators; this may be an important cause of their confidence. The confidence of the confabulator and the confidence of the sociopath are the same, in the views described in chapters 4 and 5. Perhaps because they have a mind-reading deficit they both lack the doubt that is placed in our minds when we know that at least one of our listeners disbelieves what we are saying.

According to this criterion, people with obsessive-compulsive disorder and those with hypochondria are not confabulatory. People with OCD, for instance, are not certain or confident that they need to wash their hands. The person with OCD genuinely *knows* that she does not need to wash her hands, or that she just locked the door. People with hypochondria are likewise not confident that they are ill. They are *afraid* that they are ill. Being afraid that *x* is true and being confident that *x* is true are quite different. Another related difference between confabulators and people with OCD or hypochondria is that in the confabulators there is a blunting or total lack of emotional (or at least autonomic) response, while OCD and hypochondria are accompanied by aversive, negative emotions. These emotional accompaniments are clearest in the case of OCD, which is what makes it so unpleasant, but hypochondria is also unpleasant for the person.

The following candidate criterion is redundant, but worth mentioning. *Jan does not intend to deceive.* This is implied independently by criterion 2 (*Jan believes that p*), and criterion 4 (*Jan does not know that her thought is ill-grounded*). Sociopaths are problematic in this regard because they seem initially to intend to deceive, but in the course of speaking show genuine belief in what they are saying (or perhaps they are merely able to suspend disbelief). They make an initial conscious decision to deceive, then while they are executing the deception operate in a more automatic, less conscious mode.

There are also what we might call adjunctive features of confabulation. They frequently occur, but are not as crucial as the core features. Among these are *that p is desirable for Jan.* Often confabulators claim states of affairs that are desirable for them, especially in their current condition: "I can move my arm," "I can remember," etc. But there also seem to be many cases where undesirable states of affairs are claimed to be the case: "My wife is an imposter" (Capgras' syndrome); "I am dead" (Cotard's syndrome).

Clinical Confabulation
One benefit of the approach used here is that it allows a clear way to identify a given claim as a confabulation independently of the notion of a confabulatory person. There is room in this approach for the idea that a normal person might on occasion confabulate, for instance by reporting a false memory, without being generally confabulatory. His or her condition is acute rather than chronic. We should reserve the term *confabulatory person* for people who have a continual tendency to unknowingly make ill-grounded claims. This definition can also be augmented in different ways to fulfill the many other functions required of a definition of confabulation. The basic six-part definition is meant to apply to all cases of confabulation as they occur among patients in the clinic, and in their milder

varieties, as they occur in normal people. Cases of confabulation seen in the clinic contain additional features, however. A further clause can be added for clinical confabulators.

7. Jan should know that not *p*   A person should know whether he can move his arm or not, whether he can see or not. Even the patients admit that they should know about the state of their bodies. A denial patient of Weinstein and Kahn (1955) said, "If I'd been wounded, I would know it." There is a problem with adding this to the primary definition, though. It would automatically rule out someone who claims all sorts of knowledge about God or some other topic about which little is really known, as a case of confabulation. Consider the person who falls within the normal range, but believes he has special knowledge of religious matters (Socrates' foil Euthyphro comes to mind). When he relates this knowledge, is he confabulating? He shares the patient's tendency to give voice to suspect thoughts, as opposed to the normal person who might do it briefly, or in a limited way. However, it seems wrong to say that this person should know that his claims are false. Rather, he should know that he does not know, i.e., he should know that his thoughts are ill-grounded (criterion 5). He is a normal confabulator, not a clinical-level confabulator.

Anosognosia
Neglect is a space-representing or -attending deficit and checking failure that leads to denials, which are confabulations. Denial of a disability itself occurs when the claim that *p* is such a denial, for instance the Anton's syndrome patient's claim that he can see. The definition needs to be relative to a specific disability, *d*, since there is abundant evidence that confabulators will admit some disabilities while denying others (see chapter 6). We can add three other conditions for anosognosia.

8. Jan has a disability, *d*   This feature explains why criterion 3 applies (*Jan's thought is ill-grounded*). The disability (or the damage that caused it) causes the ill-grounded thought to be produced.
9. Jan does not know that she has *d*   This feature explains why criterion 4 applies (*Jan does not know that her thought is ill-grounded*). If she knew that she had this disability, and understood its nature, she would know that her belief is ill-grounded.
10. Jan should know that she has *d*   Her problem is severe and apparent enough that it should be noticed.

Capturing the Spontaneous versus Provoked Distinction
Setting up the definition in this way also allows a simple addition that can satisfy many of the same needs that required the postulation of two grades

of confabulation, roughly a severe and a mild variety. There is an important difference between the Korsakoff's patient who believes he went to a business meeting this morning, and the misidentification patient who believes three different people are his wife. Kraepelin (1904) suggested a distinction between fantastic and simple confabulations. Berlyne (1972) offered the fantastic versus momentary distinction, and Kopelman's (1987) distinction currently is the best known—the spontaneous versus provoked distinction. While it is true that patients who spontaneously voice confabulations are usually more seriously ill than those who only confabulate when asked certain questions, there are too many exceptions to this for the distinction to be useful. Also, the fact that something is said in response to a question does not determine the degree of severity it indicates. There is a way to capture the fundamental mild–severe dichotomy without confounding the process of confabulation with our method of studying it: the plausible–implausible distinction.

A plausible confabulation can be captured with condition

8a. That p is plausible   One way to explain what is meant by "plausible" is to ask whether one would believe confabulators if they were encountered outside the clinic. If you were talking to a stranger on an airplane flight, would you believe his confabulations? What the Korsakoff's patient says about going to a business meeting is plausible. What the Capgras' patient says about his mother is implausible.

An implausible confabulation can be captured with condition

8b. That p is implausible   This implies that there is additional brain damage that is causing an implausible (as well as ill-grounded) thought to be generated. In this case, p itself may also be classified as a delusion. Confabulating and being delusional are different categories. One can be delusional and not confabulate if one avoids asserting and defending one's delusional thoughts. There do seem to be clear cases where confabulators are also delusional, as in Capgras' syndrome.

Two Types of Epistemic Theory

We could call our definition of confabulation an epistemic grounding theory. Such theories are a subclass of a more general type of definition based on knowledge. There are two basic reasons that contradict the claim that someone knows that p: (1) p is false, and (2) the person's belief that p is ill-grounded. This implies that there is a basic epistemic theory, and two more specific theories—a falsity theory and the grounding theory described earlier.

In the basic epistemic theory,

Jan confabulates if and only if:
1. Jan claims that p.
2. Jan believes that p.
3. Jan does not know that p.
4. Jan does not know that she does not know that p.
5. Jan should know that she does not know that p.
6. Jan is confident that p.

In the epistemic falsity approach,

Jan confabulates if and only if:
1. Jan claims that p.
2. Jan believes that p.
3. That p is false.
4. Jan does not know that p is false.
5. Jan should know that p is false.
6. Jan is confident that p.

Given that several early writers have claimed that falsity is a necessary condition for confabulation, what is wrong with the falsity approach? One way to arrive logically at an epistemic grounding theory of confabulation is to begin by directing one's attention to the truth values of a patient's claims. If we say that confabulations are by definition false utterances, when a highly confabulatory patient guesses correctly that it is Tuesday when we ask him what day it is, he is not confabulating. This indicates that the ill-groundedness may be more important than the falsity. Otherwise, without other criteria, all expressions of false belief are confabulations.

Conversely, if we imagine someone who is not at all confabulatory, but who through no fault of his own says something false, it seems wrong to claim that he confabulated. The problem is not the falsity itself, but that the claims are being produced by a malfunctioning cognitive system, which is producing ill-grounded thoughts. If we did not know that the claims are produced this way, it might cause us to give less weight to the ill-groundedness component of knowledge. However, we know that the confabulator's belief is ill-grounded, and we know why: He has sustained serious brain damage. A similar dynamic plays out in the case of lying. If someone intending to lie unknowingly speaks the truth, he may escape the charge of lying on that occasion, on a technicality. However, we still distrust him and consider him a liar because he misrepresented what he (falsely) believed. In this respect, "to confabulate" is closer to "to make up." If I make up a description of what happened, even if it turns out to be true, it is still correct to say that I made up the description. This issue is raised again when we discuss whether falsity should be a criterion for confabulation.

Since the claim itself need not be in the form of a sentence expressing that p (it may be in the form of a picture, a hand signal, etc.) we may also allow a more general definition designed to apply to all such claims or acts of representation.

Jan confabulates in representing that r if and only if:
1. Jan represents that r.
2. r is accurate, for Jan.
3. r is ill-grounded.
4. Jan does not know that r is ill-grounded.
5. Jan should know that r is ill-grounded.
6. Jan is confident that r is accurate.

8.5   Other Candidate Criteria and Conceptions

In chapter 1, I introduced three different conceptions of confabulation: the mnemonic, the linguistic, and the epistemic conception I have been defending here. The driving force behind the mnemonic conception is a candidate for inclusion in the definition of confabulation: the patient has a memory problem. The linguistic conception also hangs its hopes on the inclusion of a certain criterion: Confabulations are expressed verbally, in natural language. The falsity criterion and the verbal criterion support each other in the linguistic conception. If the emphasis is moved onto the act of speaking itself, it is removed from what is going on in the mind of the person making the statement, and the defect must be attributed to the statement itself.

These candidate criteria include the following: (1) The claim is false or incorrect. (2) The claim is in linguistic form. (3) The patient has a memory defect. (4) The patient intends to deceive. In this section, I offer reasons against adding each of these to the definition of confabulation.

Definitions That Make Falsity a Necessary Condition
Feinberg defines a confabulation as "an erroneous yet unintentional false statement" (2001, 55). Talland says that a confabulation is a "false verbal statement about facts" (1961, 362). Berlyne's classic definition of confabulation is that it is "a falsification of memory occurring in clear consciousness in association with an organically derived amnesia" (1972, 38). In a later book, Talland adds several modifiers, defining a confabulation as a "factually incorrect verbal statement—other than intentional deception,

fantastic fabrication, wild guess, gibberish, or delusion" (1965, 56). Such definitions are far too broad as they stand, however, because they make almost every false statement into a confabulation.

I noted earlier that the best objection to the falsity criterion is that a confabulator might say something true as a matter of luck. This should still be regarded as a confabulation because its ill-groundedness is more important than its truth or falsity. We are concerned with the degree of credibility we can expect from the person in question. A patient who gets a question right after supplying wrong answers to the previous six has not miraculously stopped confabulating. Confabulator's claims have a low probability of being true because of brain malfunction. We expect each other to exercise a certain amount of control over the truth value of the claims we make. Abraham Lincoln makes this point in a letter to the editor of a paper that had published accusations that he was a "religious scoffer," no doubt a serious charge in Illinois in the 1840s. "I believe it is an established maxim in morals," writes Lincoln, "that he who makes an assertion without knowing whether it is true or false, is guilty of a falsehood; and the accidental truth of the assertion, does not justify or excuse him" (1846/ 1953, 384).

The linguistic conception of confabulation needs a falsity criterion since its emphasis is much more on the linguistic act of confabulation than on the (damaged) epistemic processes leading up to it. Berrios (1998) contends that the partial falsity present in many confabulations of Korsakoff's patients makes determining falsity a difficult and perhaps arbitrary matter. In a case where the content of the memory itself is correct, but the represented event is displaced in time (i.e., usually an old memory is mistaken for a recent one), is the memory report false or not? Strictly speaking, it is false, but to say this does not distinguish it from a less plausible confabulation where a patient describes an event that never happened. Berrios concludes from the partial falsity problem that "if truth value is not always crucial, there are cases where even a *true statement* may be considered as confabulatory. These cases suggest that the real issue is the evidential support on which the subject bases his claim" (1998, 234). However, these considerations do not make it into Berrios's final definition of confabulation, except to cause him to put quotes around the word "untrue": Confabulations are "'untrue' utterances of subjects with memory impairment" (1998, 237). (Berrios describes a second type of confabulation that involves "fantastic content" and "great conviction," which is similar to the notion of spontaneous confabulation; hence he is not committed to the claim that all confabulators have a memory problem.)

The definition offered by Moscovitch and Melo suggests a way to deal with the problem of partial falsity. They say that confabulations are "statements or actions that involve unintentional but obvious distortions"

(1997, 1018). The notion of a distortion is unclear, though, and is still subject to the objection that a person might make a distortion-free claim out of luck. A second problem with the Moscovitch and Melo definition has to do with the requirement that the distortions be "obvious." To a listener who does not know that a person is confabulatory, what many confabulators say sounds quite plausible. Whether or not someone finds a claim obviously false can also be doubly dissociated from the question of whether it is ill-grounded. In his review on the cognitive neuroscience of confabulation, DeLuca employs the Moscovitch and Melo definition because it is "the most general yet descriptive," but he seems uncomfortable with the obviousness criterion—he says a few sentences later that "confabulation can be subtle, nonbizarre" (2001, 120).

When we call someone confabulatory because we notice that the first memory claim he made is false, followed by a second or third false claim, we are assessing the credibility of the patient. One's credibility is not automatically zero after the first false claim, but it drops in a continuous manner toward zero with further false claims. Once a person has produced enough false claims, we regard him as confabulatory. This implies that we are using the falsity as an indicator of something more crucial—the groundedness of the claims. The idea that confabulations are fictional might be of some use here, since works of fiction often contain true statements, for instance when they contain actual historical characters.

Definitions According to Which Confabulations Are Necessarily Verbal

Recall that Feinberg defines a confabulation as "an erroneous yet unintentional false *statement*" (2001, 55). Wegner defines confabulations as "convenient *stories* made up to fit the moment" (2002, 47). Myslobodsky and Hicks begin their definition of confabulation by stating that "confabulations can be defined as *verbalizations* containing various degrees of distortions of the target material" (1994, 226) (my italics). A linguistic theory might add a "saying" requirement to the requirement that the patient claims that p: Jan claims that p by saying that p. Or it might add a criterion mentioning that p is asserted. A weak linguistic theory might just use the "say" criterion, without a belief criterion, to avoid the question of whether the patient believes what he says. You can say that "Jack won the lottery," and no matter what is in your head, you can say something true; there are fewer intentional requirements present in saying than in asserting.

After describing the experiment by Bender et al. (1916) in which incorrect additions to drawings of meaningless patterns made from memory were referred to as confabulations, Talland says, "customarily confabulation refers to a certain class of verbal statements, and there seems to be no compelling reason for stretching the application of the term beyond the confines of verbal behavior; its meaning is confusing enough within that area" (1965, 41–42). While we can all agree with the latter statement, we need not draw such a line out of convention. We might discover empirically that some sort of "say" criterion is truly required by the data; if we found, for example, that the sort of overreaching described in the epistemic theory only happened when a person spoke and did not occur when a person made claims in other ways, by pointing, by drawing, etc. The data do not bear this out, however; in chapters 6 and 7 we reviewed numerous studies showing that confabulatory patients confabulate in nonverbal ways also.

Definitions That Make a Memory Defect Necessary
Mercer et al. state flatly that "a necessary (though not sufficient) prerequisite for confabulation is impaired memory function" (1977, 433). As indicated in chapter 1, the American Psychiatric Association defines confabulation as "the recitation of imaginary events to fill in gaps in memory" (1994, 157). Myslobodsky and Hicks build in a criterion reminiscent of Schnider's theory (Schnider et al. 2000) about the role of the orbitofrontal cortex in keeping irrelevant memories at bay. They say that confabulation involves "unusual avidity to irrelevant associations in the realm of memory" (1994, 226). Berlyne's definition contains both a falsity criterion and a memory criterion: "Confabulation can be defined as 'a falsification of memory occurring in clear consciousness in association with an organically derived amnesia'" (1972, 38). Although he mentions that confabulation has been applied to anosognosic patients without memory deficits as well as to normal people, Berlyne presents no arguments against the claim that these are genuine cases of confabulation. After endorsing the Moscovitch and Melo (1997) definition, which is neutral with regard to memory, DeLuca lapses into assuming that confabulation must involve memory: "At a minimum, confabulation involves both distortions of content and temporal context. That is, confabulatory recollections frequently include additions, distortions, or elaborations of events that either actually or plausibly occurred" (2001, 120).

Again, we need not rely on mere convention; empirical discoveries can help here. I noted in chapter 3 that confabulation and memory deficits can be doubly dissociated in both Korsakoff's patients and patients with ACoA aneurysms. However, there is also room to argue that memory problems are at the root of many if not all of the confabulation syndromes. When we surveyed the confabulation syndromes, patients other than those with Korsakoff's syndrome or an aneurysm of the anterior communicating artery on occasion showed memory problems, but they typically were not of the same kind. Even within the memory confabulation syndromes, one finds different areas of memory (episodic versus semantic) compromised in different patients. Depending on how broadly one defines memory, one might find a conceivable way to claim that split-brain patients, anosognosic patients, and misidentification patients have a memory problem, but only in very short-term memory. To do so, however, is to stretch the concept of memory beyond limits that would be useful in characterizing the phenomena of confabulation. There are also perceptual confabulations where invoking the idea of memory seems pointless; for example, when an Anton's patient "describes" what he sees.

Definitions That Contain an Intentional Deception Condition
The earlier statements in defense of the claim that confabulators believe what they say also applies here, of course. Several of the early definitions left it ambiguous as to whether or not confabulation involves intent to deceive. These can be traced back to Bonhoeffer's notion of "confabulations of embarrassment," and its implication that embarrassment provided the patient a motive to consciously deceive the listener. But as Whitlock (1981) points out, the idea that the patient is embarrassed about a bad memory implies that he is aware his memory is bad, which is contrary to what we know. I also noted in chapter 5 that confabulators show mind-reading deficits, and this militates against both their being embarrassed and their engaging in intentional deception, since deception often requires mind reading. Some definitions address this question directly by adding a criterion about lack of intent to deceive: "Confabulation can be broadly defined as an erroneous statement made without intent to deceive" (Feinberg and Roane 1997, 73). This eliminates liars from the set, but still, a huge number of simple mistakes by people, normal or abnormal, get included as confabulations.

Is the Wernicke's Aphasia Patient a Confabulator?
Wernicke's aphasia is a language disorder in which the patient can speak words fluently, but his or her speech includes neologisms and enough semantic errors that it often makes no sense at all. Wernicke's patients also have trouble comprehending language and typically also have trouble naming things, or even repeating phrases said to them. Here is an example of what a Wernicke's patient said when asked what brought him to the hospital: "Boy, I'm sweating, I'm nervous, you know once in a while I get caught up, I can't mention the tarripoi, a month ago, quite a little, I've done a lot well, I impose a lot, while, on the other hand, you know what I mean, I have to run around, look it over, trebbin and all that sort of stuff" (Gardner 1974, 68).

In the epistemic view outlined here, the nonsensical statements of Wernicke's aphasics would not count as confabulations, since confabulating involves claiming something to be the case, and Wernicke's patients are not making coherent claims, they are merely throwing out words. It often happens that when we are able to discern what a Wernicke's patient is attempting to say, it is correct and reasonable, and the epistemology of it is fine; the problem is the grammar and word choice, the use of nonwords, etc. These patients do not understand what they are saying, whereas the confabulators understand what they are saying. While I have urged a liberal sense of "claim" so that all types of pointing, hand signals, drawing, etc. can count as making a claim, we cannot accept that the Wernicke's patient is making a claim until we know what it is he is saying. We cannot ignore the content of the claim itself, which including Wernicke's syndrome would frequently force us to do. It may also be an important part of the dynamic of confabulation that the patient's hearers take him to be making a claim. Wernicke's patients do have unawareness, as evidenced by the fact that they do not attempt to correct what they say (Heilman, 1991), but as we have seen, unawareness may not lead to confabulation if the person does not make some sort of claim.

Talland seems to agree with this approach: "Totally incoherent talk ought probably to be attributed to delirious fantasy; utter nonsense or gibberish should certainly be excluded from the definition of confabulation" (1961, 366). If we think of confabulation merely as random, "blind" talking, then Wernicke's syndrome seems to fit. The blindness in confabulation is epistemic blindness, however, while the blindness in Wernicke's is syntactic, or semantic (in the sense of word meanings). Another reason it is different is that the errors in Wernicke's syndrome and in confabulation involve different linguistic categories. In Wernicke's the errors are syntactic and semantic, in the sense that they get word meanings wrong; in confabulation they are semantic (in the sense of truth values), pragmatic, and epistemic. The types of confabulators we have examined do not make syntactic errors. The knowledge deficits in confabulation are deficits in knowing *that*, as opposed to knowing *how*. The problem of the Wernicke's patient is that he or she no longer knows how to speak a language, but does not know this.

8.6   Epistemic Features of Confabulation

Attempts by philosophers to construct a necessary and sufficient definition for "know" date back to Plato. The task has proved difficult but informative. The orthodox definition that contemporary theorists of knowledge either defend or attempt to refute is that Jan knows that p if and only if that p is a justified, true belief of Jan's. The truth criterion seems unassailable if we accept that every proposition is either true or false. Finding a counterexample to it would involve someone knowing that p, where p is false—impossible. The belief criterion also seems quite sensible, judged by the oddness of saying Jan knows that p, but she does not believe it. The attention, then, has been toward the justification criterion.

Theories of justification are traditionally divided into internalist and externalist approaches. In an internalist account, our beliefs must be justified by internal states, "to which we have 'direct access'" (Pollock 1986, 22). My belief that my car is in the parking lot, for instance, is justified by my seeing the car in the lot. We should be able to become aware of these states easily, by merely reflecting. Externalism is the denial of internalism. According to the externalist, there are other factors important in justifying a person's beliefs that may not be accessible to him or her.

Confabulation involves epistemic failures of both internal and external types. Internally, we are concerned about the confabulator's failure to be aware of the implausibility of his or her remarks. For the anosognosic patient who denies paralysis, the experience of seeing an unmoving limb should be a mental state to which he has direct access. Confabulators often seem to believe that their confabulations are justified by other states, by alleged memories or perceptions, but typically those states are ill-grounded also. They have access to what they believe is supporting information, but it is not. Confabulators fail to live up to basic epistemic norms regarding things we all take for granted: basically sound memory, the ability to recognize people, and knowledge of the state of one's body. Externally, since we know the clinical confabulator has sustained brain damage, and we have independent evidence that this particular brain damage can cause perceptual or mnemonic problems, we tend to attribute the ill-groundedness to this cause, whether or not the patient is aware of the damage.

Criterion 5 in our definition of confabulation (*Jan should know that her thought is ill-grounded*) links internal and external considerations. Her brain damage is an external feature of Jan, but Jan should have formed an internal state, a representation to the effect that she has brain damage. Of all the ailments the human body is heir to, there is a rough division into those one should know one has, and all the rest. One should know whether one is blind, one may not know that one has a heart murmur; one should know whether one is paralyzed, one may not know whether one has osteoporosis. Since the normal processes for gaining knowledge are disrupted, confabulations are often constructed using non-normal processes. We know that the denial patient's claim that his arm is fine is ill-grounded because we know that an area crucial in the generation of those claims is destroyed. This is clearer in the case of Anton's syndrome patients. We know that a patient's belief that he sees three fingers is ill-grounded because we know he is blind.

The Background

Normally when someone makes claims about what he sees right in front of him, his ability to move his arm, or whether someone is his father, we accept it without question. It even sounds odd to say, "I believe I can see." These are the sorts of things one simply knows, we want to say. We assume they are true, and often pass our entire lives without explicitly asserting them. If asked about one of these facts, initially we might reflexively assent to these basic certainties, but what is different about us is that once a problem with them was pointed out, we would test them more carefully. Part of the problem is that normally these capacities do not malfunction—they are not normally part of the range of thoughts that we doubt. This may be one reason confabulators have trouble; we have not developed any capacities for recognizing and correcting malfunctions at such a basic level.

In his article boldly titled "Proof of an External World," G. E. Moore (1942) described holding up his hand, and said that his claim to know his hand exists is far more certain than any possible doubt about its existence. In response to this, Ludwig Wittgenstein (1969) objected that there is something peculiar about holding up one's hand and saying "I know my hand exists." What troubled Wittgenstein was that both knowledge and doubt presuppose the possibility of giving evidence. The way to give evidence for a claim is to find something more certain than the claim that supports it. However, what is more certain than that this is my hand? In cases such as this, Wittgenstein argued, doubt does not make sense. If certainties such as this are the background for thought, then they cannot be questioned or doubted, for epistemic reasons, in the same way that you just have to believe your axioms when you solve an algebra problem. Wittgenstein (1969) called these basic truths "framework facts." Searle (1983) called them "the background." In Searle's picture, the background does not consist of states such as beliefs, but rather it makes such states possible. Our relationship to a claim such as, "This is my hand" may be as close as we can come to understanding the state of mind of the confabulator. We have no real capacity to doubt a claim such as that, just as the confabulator has lost his capacity to doubt the claims occurring in the affected knowledge domain.

These considerations may also show that many of the thoughts that yield confabulations when expressed are thoughts that had to be created rather than merely remembered. The thought that I can move my left arm is not explicitly stored in semantic memory, as many of our beliefs are. Notice that many of the knowledge domains compromised in confabulators involve primarily analog or image-like representations. We normally represent our body with analog representations; visual representations are the paradigm of analog representations; episodic memories seem to be analog representations; mind reading seems to employ both conceptual and analog representations, but the ones we are most interested in seem to be more analog, i.e., representations of the moving face. This suggests that part of the dynamic of confabulation has to do with problems in the processes involved in "translating" an analog representation into a conceptual representation to be expressed (Joseph 1986; Hirstein 2000).

Risks and Rewards in Belief Formation

Once a thought is clearly formed, we need to consider whether or not we believe it. Whenever a person considers promoting a thought to the level of belief, certain risks and possible benefits must be brought into the equation. With regard to the truth of beliefs, these risks and possible rewards divide into four categories:

1. S believes that p and p is true.
2. S believes that p and p is false.
3. S does not believe that p and p is true.
4. S does not believe that p and p is false.

If your checking processes are hypoactive, the odds of you accepting p as a belief increase, especially if the facts the belief describes are in the set of facts that cannot be checked because of a brain lesion. This increases the odds that you will commit the error of believing something that is false (category 2). Sometimes it can be disastrous to believe that p when p is false, while other times it is seemingly of no consequence. These negative consequences will of course vary from person to person, and for a given person they will vary from situation to situation. Conversely, if your checking processes are hyperactive (as in OCD), it becomes less likely that you will accept a given belief. But this increases the chance that you will fail to believe that which is true (category 3).

Confabulators have too low a threshold for accepting thoughts as beliefs. The acceptance threshold cannot be an absolute probability, since it needs to be sensitive to several factors, including the number of thoughts a person generates. If you tend to generate a lot of thoughts, the number of beliefs you form per day will increase rapidly if belief formation is based on a set probability. There is another factor in the equation, however: how much time and effort will be required to increase the probability of the thought if it is too low. People do not normally consciously and intentionally set these thresholds for different beliefs. Rather, this is something brains do for their owners. It is possible, however, to explicitly and voluntarily alter one's thresholds. The scientist speaks more freely about his views when he is with friends, and is more cautious at a professional conference.

What people must be able to do when they consider a belief is sense the level of their thresholds. These levels seem to vary in a continuous manner, smoothly moving up and down, rather than in discrete jumps between a few possible values. We might also gain insight into the epistemic nature of confabulation by studying the epistemic features of its opposite, the pathological doubt in obsessive-compulsive disorder. Confabulators believe something they should doubt; people with OCD doubt something they should believe. Just as confabulators have an irrational confidence in their beliefs, OCD patients seem to have an irrational lack of confidence in theirs.

Process Reliabilism

One type of existing theory of knowledge that the notion of confabulation described here could be fitted into is called process reliabilism. A belief-producing process is reliable if it tends to produce true beliefs. Here truth enters into the picture when we flesh out justification, or groundedness. Being produced by reliable processes provides justification for a belief. Here is how Goldman (1998) details his approach to the justification criterion:

If S's belief in p at t results ("immediately") from a belief-independent process that is reliable, then S's belief in p at t is justified.

If S's belief in p at t results from a reliable cognitive process, and there is no reliable or conditionally reliable process available to S which, had it been used by S in addition to the process actually used, would have resulted in S's not believing p, then S's belief in p at t is justified.

Confabulating patients violate both the first *and* the second clause. In the case of Korsakoff's syndrome, memory is no longer a reliable process; in Anton's patients, hallucinations are produced by a process that must be judged highly unreliable if it is supposed to be providing perceptual information. Given this initial problem, a second line of defense also fails; the patient is unable to notice and correct the ill-grounded representations. This second problem seems to cause confabulators to violate Goldman's second condition. The orbitofrontal processes that assess representations receive candidate beliefs as input, but are not operating in confabulation. If they were operating, this would be enough to cause the patient to reject the belief in question.

In cases of confabulation, it might be that whatever process produces the claim that the patient is fine in anosognosia is in fact a normally reliable process. The problem in this case is that it is out of its domain. Goldman's second condition would catch this though, because there is a better process involving the right inferior parietal cortex that would have resulted in the patient not believing that he is fine.

Thinking of reliability in terms of probability of being true has two things to recommend it here. First, it brings truth into the account, and this honors our intuition that truth should play some role in the explanation of confabulation. A second reason, specific to confabulation, is that it may capture a crucial failure in confabulation: the ability to assess the probability of a thought being true. When something causes us to distrust either memory or one of the senses, we revise the truth probability attached to the representations delivered from the suspect process. It would be counterproductive and impossible to completely distrust either memory or the senses. The failure to satisfy criterion 4 (*Jan does not know that her belief is ill-grounded*) is relevant here: Jan fails to revise the probability of her belief downward when she has sure signs the belief is defective.

Damage to a person's orbitofrontal checking and inhibiting processes lowers the probability of that person's beliefs being true. Perhaps one interesting thing confabulation shows here is that there are processes relevant to the reliability of a person's beliefs that are not themselves involved in producing beliefs. The function of many of the checking processes is more to eliminate candidate beliefs than to produce beliefs. They aid the cognitive system as a whole in producing a set of beliefs that has a higher average probability of being true. Note that ill-groundedness is more complicated than just having a truth probability below a set threshold. The threshold depends on the situation the person is in, the content of the belief involved, and the knowledge domain involved. The confabulators' contexts are those that normally involve very high probability claims.

One criticism of externalism is that there are logically possible situations in which people's minds are filled with false beliefs as a result of purely external considerations, yet we have an intuition that the beliefs of such people are justified. They are justified in the sense that the people are epistemically blameless—there is nothing they can really do to know their true condition. As we saw, some researchers think that confabulators are in a similar state, in that they are getting false memories or perceptions, but their belief-forming processes are still functioning correctly. In the case of the confabulators, there should still be other checking processes that can indicate problems with ill-grounded beliefs and if necessary reject them. Other criticisms of the reliabilist perspective center on its heavy reliance on truth. We do want our beliefs to be true, but we also want them to be useful for us. Memorizing the phone book would be a good technique for maximizing the number of one's true beliefs, but of almost zero use. We need answers to certain crucial questions. These two demands—the demand for truth and the demand for usefulness—can come into conflict.

8.7 Knowing That We Do Not Know

Confabulators don't know that they don't know what they claim. Sometimes no real harm is done when people fall in this category. Suppose I don't know about modern dance, but think I do. I say lots of silly, incorrect things and look like an oaf. The problem with confabulation, though, is that the items of knowledge in question are much more vital to the patients' belief systems: knowledge about the identities of people close to them, knowledge about the state of the body, and so on.

When you know that you don't know the answer to a question, and someone tells you the answer, you are in a position to appreciate and accept it. When you don't know that you don't know the answer, and someone tells you the answer, you may not be in the proper position to accept it. What does it mean to know what you don't know? Here is one interpretation: There are many different areas of knowledge. Ideally, we should represent how much we know in each area. It might also be easier and more natural to represent our knowledge compared with that of other people, rather than attempting some more absolute gauging of it. One way to know that you don't know about a topic is to converse with someone who really does know about that topic, and learn how deficient your knowledge is. Another way to know that you don't know the answer to a question is to one-by-one reject the possible answers that occur to you. If one can use, either intentionally or nonintentionally, frontal processes to eliminate possible answers, this can exhaust you and make you admit ignorance.

"Don't Know" Tests

In chapter 3, we saw that some researchers (Mercer et al. 1977; Dalla Barba et al. 1990) used "Don't Know" tests to measure confabulation. Typically these contain questions of varying degrees of difficulty, with the most difficult questions designed to be so hard that any normal person would admit ignorance. Suppose you are asked ten difficult questions, to which you must answer either "yes," "no," or "I don't know." The ideal here is to get all ten questions right of course. But given that you are fallible, there are going to be some things you don't know, so the next best thing to giving a correct answer is to avoid giving wrong answers. You can achieve this by being aware of your epistemic limitations and answering "I don't know," when you don't have clear, warranted knowledge of the answer. Given your fallibility, your best strategy is to answer the ones you know and abstain from answering the others by saying, "I don't know." In most situations, it is better to say, "I don't know" than to give a wrong answer.

In both the laboratory and the clinic, an operational definition of amount or degree of confabulation would be useful for comparisons between different types of confabulating patients, to discern the severity of confabulation, and to track changes in patients over time. If we accept that confabulation is at least closely connected to the failure to respond, "I don't know," simple tests of the sort described here can be devised. The raw number or percentage of false responses on such tests is not a useful statistic, since that may reflect only the difficulty of the task. Similarly, the number or percentage of "Don't know" responses alone is not a good measure, since these need to be placed in a context of the total number of wrong responses and the total number of questions. Confabulation is not merely the failure to say "I don't know," it is the failure to say "I don't know" rather than give a false response. Computing the ratio of "Don't know" to false responses gives the desired result, in percentage form, as follows: the confabulation percentage is the number of wrong answers divided by wrong + don't know answers. Someone who is confabulatory and never responds "I don't know" will score 100 percent; someone who eliminates wrong responses by answering "I don't know" rather than getting the question wrong will score 0 percent.

This approach also captures the potential risks and benefits of attempting an answer as opposed to admitting ignorance. In many situations, giving a quick, confident answer carries rewards in that one is seen as knowledgeable and decisive. This reward must be weighed, however, against the possibility that the answer is wrong and that one's listeners will discover this fact, possibly at great cost to themselves. What normally follows in that event is a large devaluation in the credibility of the source of the information (although various things can happen to mitigate the size of this devaluation, e.g., the source convinces his listeners that this was a rare exception). Admitting ignorance has an initial disincentive in that it makes one appear limited, uninformed, indecisive, or all three, but it avoids the potential negative results brought about by later disconfirmation. The acumen of the audience and the speaker's perception of it also play an important role in these "Don't know" strategies. Clinicians often remark that some of their Korsakoff's patients are so convincing that if they were speaking to another person outside the clinic who didn't know of their condition, the other person would believe their confabulations. However, in the clinic, they should know that the doctor they are speaking to knows that what they are saying is false; again, they have a mind-reading problem.

Don't Know tests have their weaknesses, however. The Fregoli's syndrome patient of Box et al. (1999) made only one confabulatory error on Dalla Barba's confabulation battery. Perhaps this is because the battery does not tap the specific knowledge deficit of misidentification patients, which may be in the mind-reading domain. The sort of memory for pictures task Gazzaniga and his colleagues used (see section 7.4) is important also, since it is an analog task, and as we noted earlier, damage in confabulation frequently seems to be to analog rather than conceptual processing streams. Another objection to Don't Know tests is that "I don't know" would be a strange answer to "Can you move your left arm?" It would be odd to answer "I don't know" to any question about the background of our knowledge, although with each one we can imagine a special situation in which it would be a reasonable question; for example, after you come out of a coma in the hospital following a car accident, you are asked, "Do you remember what happened?" and "Can you move your arms?"

8.8   Conclusion

The idea that confabulation is an epistemic notion is implicitly present throughout the literature on confabulation syndromes. It may be that an easy reliance on letting the claim that confabulations are false handle all the epistemic factors caused those factors to be missed. But as we have seen, falsity alone says nothing about what caused the falsity, or why it is likely to happen again. The early literature contains several examples of "confabulation" being used in an epistemic sense. Geschwind, for instance, says: "The same patient gave confabulatory responses to visual field testing when there was no stimulus in the field" (1965, 597). From the way Geschwind uses "confabulation" and "confabulatory," he seems to, consciously or not, hold an epistemic theory. His uses of these words make this clear: "I have seen a confused patient who gave confabulatory responses when asked to name objects held in his hand" (1965, 597). He says that patients with right temporal injuries were different from those with left temporal injuries in a test of picture memory, in that "the number of correct 'yes' responses was almost the same for both groups, but ... nearly all the difference in scores was due to the large number of false 'yes' responses by the right temporal group," and suggests that this is "another example of confabulatory response resulting from a lesion of right hemisphere association cortex rather than a special deficit of the right hemisphere in the retention of unfamiliar material" (Geschwind 1965, 602). This is exactly the sort of behavior that would yield one a high confabulation percentage in the technique outlined earlier.

Johnson and her fellow researchers are also on target from our point of view when they say, "an amnesic is us when we cannot remember and know we can't. A confabulating patient is us when we do not remember accurately and don't know that we don't" (2000, 360). Some of the existing definitions also appear to have criteria that relate to epistemic features of confabulation, such as the definition from Heilman et al.: "Confabulation is the production of responses that are *not rooted in reality*" (1998, 1906) (my italics). This seems to intend a falsity criterion, although one could argue that the idea that a response should be "rooted in reality" includes the idea of sound perceptual and other epistemic processes behind it, i.e., it is similar to being ill-grounded.

As we saw, the DSM-IV (1994, 157) defines confabulation as "the recitation of *imaginary* events to fill in gaps in memory." Bleuler said that confabulations are "free *inventions* which are taken as experiences" (1924, 107) (my italics). Cameron and Margaret say that confabulation involves the "*making up* of a likely story" (1951, 537) (my italics). When Wegner (2002) says that confabulations are stories made up to fit the moment, the notion of making something up seems to imply that the made-up claim has epistemic problems. Indeed, why not just say that "to confabulate" means "to make up"? For example, "He confabulated memories" means "He made up memories." One serious disanalogy, though, is that we do not tend to use "make up" for nonintentional creations.

Now that we have a conception of confabulation, along with a more exact definition, I want to make what I think is one of the most immediate and obvious applications of this information. The denial that some patients engage in seems similar to a milder form of denial we all engage in, known as self-deception. In chapter 9, I locate the phenomena of self-deception in the set of data we have established so far. This includes relating what we know about confabulation, sociopathy, mind reading, and obsessive-compulsive disorder to everyday self-deception.

9   Self-Deception

If the propensity for self-deception is frequently a curse, this curse also seems to be a source of the bets we make about our unknowable future, and on which depend not only the greatest creative achievements of humanity, but also the savage, inexplicable hope which feeds us and sustains us in our lives.
—Eduardo Gianetti (1997)

9.1   Confabulation: Clues to Self-Deception

There is a tension deep in our nature that makes us hide things from ourselves. We don't let ourselves think certain thoughts clearly because of the emotions we know they arouse. Consider the man whose wife is having an affair. All the signs are there: She comes home late from work, disheveled and glowing, both figuratively and literally. But when she is late yet again, he tells himself, "She would never cheat on me. There are all sorts of things that can make a person late." He has a self-deceptive belief that his wife is faithful to him. We are often deceived about how good we are at something. A nice demonstration of this is the blatant conflict contained in the following fact: 94 percent of educators describe themselves as better at their job than their average colleague (Gilovich, 1991), which would, one supposes, yield a self-deception rate of 44 percent. Using the term *deceive* to describe this phenomenon creates a riddle, however. If I deceive you into believing that p, (normally) contained in this claim is the idea that I believe that not p. Applied to self-deception, then, when I deceive myself, I cause myself to believe that p, while at the same time believing that not-p!

There is no need to invent an epistemic theory of self-deception to show how it connects to the epistemic definition of confabulation outlined in chapter 8 because such theories already exist. "Self-deception is ... a form of self-induced weakness of the warrant" according to Donald Davidson (1986, 89). "Weakness of the warrant can occur only when a person has evidence both for and against a hypothesis; the person judges that relative to all the evidence available to him, the hypothesis is more probable than not, yet he does not accept the hypothesis (or the strength of his belief in the hypothesis is less than the strength of his belief in the negation of the hypothesis)" (Davidson, 1986, 89). In both confabulation and self-deception, the problem is that the warrant for the crucial thought is weak, for certain reasons. The self-deceived person's belief is not based on the most reliable belief-forming processes his brain could supply if it were not in a self-deceptive mode. The clinical confabulator's belief is not based on the most reliable belief-forming processes his brain could supply if it were not damaged. The primary difference between the two phenomena is that the self-deceived person often has a greater ability to access the information or processes that are needed to show that his belief is ill-grounded, whereas in the confabulator, the normal routes of access have been destroyed.

Confabulation patients, especially those exhibiting denial of illness, seem to be engaged in a form of dense and completely successful self-deception. In contrast with them, the greater access that the self-deceived person has to the disconfirming information often produces a tension within his psyche. Thus, while criteria 1 to 5 of the definition of confabulation offered in chapter 8 can fit certain cases of self-deception, where the expressed belief is arrived at through self-deception, criterion 6 (*Jan is confident that p*) does not apply to self-deceived people who experience tension. The self-deceived normal person who is experiencing tension does not make his claims with the serene confidence of the confabulator. The absence of conflict in someone may be enough to prevent us from thinking of that person as self-deceived. It is enough for Davidson: "The existence of conflict is a necessary condition of both forms of irrationality [self-deception and weakness of the will], and may in some cases be a cause of the lapse" (1985, 82). If Davidson is right about the necessity of conflict, this would sharply distinguish self-deception from clinical confabulation, where there is no conflict because one of the normal participants in that conflict is not present.

There do seem to be cases of self-deception with little or no tension, though (see also Mele, 2001). The people who were self-deceived about being better than average, for instance, are not experiencing any tension about this. Many of them do not have the standards to see that they are, alas, submediocre. They do not know enough about the competencies required, their own performance, or the population involved to answer correctly. These are epistemic standards. These self-deceived people are closer, then, to confabulators and sociopaths in that they are tension free.

In brief, the connection between self-deception and confabulation may be as follows. There is a kind of looseness in the cognitive apparatus between a given thought and the frontal processes capable of checking it before allowing it to become a belief. In the case of self-deception, some process is capable of keeping the checking processes away from the thought that needs to be checked. In confabulatory neurological patients, the checking processes themselves are damaged, or in some more permanent and concrete way kept from evaluating the candidate belief. One can then frame the question of whether or not the self-deception is intentional in terms of whether the agent has voluntary control over these processes. There are two places where voluntary control might operate. Either the thought could be kept away from checking processes, or the checking processes could be kept away from the thought. It is possible that both of these can be done intentionally or nonintentionally.

In cases where there is conflict, as Losonsky (1997) puts it, self-deceived people possess evidence against their self-deceived belief that is "active" in their cognitive architecture, and this manifests itself in the person's occasional tension or nagging doubts. The self-deceived person has the processes needed to improve his or her belief state but does not employ them. The man whose wife is having an affair has a knowledge deficit, a blockage based on an emotional reaction. The self-deceived person fails to take normal, easy-to-employ, and obviously applicable epistemic checking measures. Clinical confabulators, in contrast, no longer have these checking processes. Sociopaths fall between confabulators and self-deceived normal people. If they possess these checking processes, the processes seem to lack the emotional force needed to inhibit unwise or hurtful actions. People with obsessive-compulsive disorder are the opposite. In their minds, the thought that they need to wash their hands, for instance, is powerful enough to interrupt the flow of other thoughts and force the person to wash his or her hands. In this view, their tension level is high because of the high activity of the frontal checking processes and their autonomic connections. All of this suggests a continuum on which these syndromes and their degrees of tension can be placed:

Clinical confabulator
Sociopath
Self-deceived normal person without tension
Normal confabulator
Neutral normal person
Self-deceived normal person with tension
Lying normal person
Obsessive-compulsive normal person
Clinical OCD sufferer

At the top end of the continuum, confabulators experience no tension at all when they make their ill-grounded claims. Tension steadily increases as we move down, peaking in the case of the person with severe obsessive-compulsive disorder and unbearable tension. The notion of tension is somewhat vague, but the findings catalogued in chapter 4 indicate that it corresponds to the activity level in the medial portions of the orbito-frontal cortex and their connections to the autonomic system. We might also describe this continuum as running from extreme certainty at the top to extreme doubt at the bottom. In general, the difference between a confabulatory neurological patient and a normal self-deceived person is that the normal person can experience a sense of tension because of the conflicting information contained in his mind. The neurological patients, on the other hand, are completely at ease and speak falsities with a felicity even a sociopath would admire, because their minds have been rendered unable to dislodge their ill-grounded thoughts and beliefs.

If it is true that the processes damaged in confabulating patients are those producing the tension in self-deceived people, we ought to be able to learn something about self-deception by studying confabulation. Understanding how self-deception relates to confabulation can also help us answer the recurring question in the philosophical and psychological literature of whether self-deception in some sense involves the deception of one agent by another.

Another piece of what we have learned about confabulation can be applied here: the distinction between the two phases of confabulation—the production of the ill-grounded thought and the failure to reject it. Self-deception has both phases of error also; an ill-grounded belief is generated, but then it is protected and allowed to remain. The metaphor of protection can apply in various ways. In some cases, a person intentionally does not allow himself to think about, remember, or perceive certain facts. A second type occurs when a self-deceived person intentionally does not use certain checking procedures.

Self-deceived people with tension tend to dwell on neither the feared thought ("She's having an affair") nor the self-deceptive thought ("She's faithful"). One way to protect certain beliefs from being dislodged is never to think them through clearly and vividly. If the self-deceived person thinks the feared thought clearly, if it exists as a detailed and well-formed representation in his brain, it will produce a strong negative emotion, and it may well initiate procedures to stop doing what he is doing and confront the negative thought.

Once the human brain initiated longer perception-action cycles during the course of our evolution, we ceased to be constantly affected by our environment, and developed the ability to attend to certain chains of thought for long periods of time, far longer than other animals. This can only be done if the attention-demanding impulses from all the things that need to be attended to can be held at bay. As Fingarette (1998) has pointed out, and as Schnider (2001) and others noted, much of the work of the brain goes into inhibiting representations from interfering with the dominant thought processes. This necessity has given the brain the power to deactivate representations or direct attention away from them in various ways. These functions might also be part of the mechanism behind certain types of self-deception.

The philosophical work on self-deception consists primarily of devising accounts of the mind (including sometimes accounts of "the self") or of belief that are able to explain how self-deception occurs. Perhaps the most natural way to approach self-deception is to treat it as analogous to normal interpersonal deception, by having one agent deceive another, all within one person's mind. These accounts of irrational behavior go back to Plato and Freud. Pears's (1984) theory is the most recent and thoroughly developed such approach. In the case of the self-deceiving husband, Pears would say that the there is in his mind a cautionary belief that it is irrational to believe that his wife is not having an affair. But this belief does not interact with the belief that his wife is not having an affair because it is contained within a subsystem, "a separate center of agency within the whole person" (Pears 1984, 87). One objection to Pears's theory is that it describes, not self-deception, but the deception of one agent by another. It also allows several agents in the mind to have their own thoughts, beliefs, etc., a move that worries some writers.

Here, as before, we need to be clear whether we are examining the current meaning of the term *self-deception* or the referent of that term, the phenomena that generated all the evidence we have, whether or not they fit our concept. As with confabulation, we also need to be open to the possibility that our current concept of self-deception may not describe a single phenomenon; we may be mixing together distinct brain processes and malfunctions. One problem with attempting to solve the problem of self-deception by working strictly at the conceptual level, as philosophers typically do, is that our folk mental concepts are flexible enough to support inconsistent hypotheses. One of the most serious flaws of folk psychology is that it overgenerates predictions. It can predict and explain why someone did what she did, or why in identical circumstances, she did not do it, with seemingly equal credibility. Another difficulty here is that there is a family of different types of self-deception, and specific definitions of self-deception can only capture paradigmatic cases, or at best some subset of the many cases that actually occur. Davidson provides a nice inventory of the phenomena: "Self-deception comes in many grades, from ordinary dreams through half-directed daydreams to outright hallucinations, from normal imagining of consequences of pondered actions to psychotic delusions, from harmless wishful thinking to elaborately self-induced error" (1998, 18). If the phenomena of self-deception are unified, and are due to definite brain processes, this is one way to affirm the explanatory value and legitimacy of the concept. Mele argues similarly that "the issue is essentially about explanation, not about alleged conceptual truths" (2001, 10). "We postulate self-deception," he says, "in particular cases to explain data: for example, the fact that there are excellent grounds for holding that S believes that p despite its being the case that evidence S possesses makes it quite likely that $\sim p$" (2001, 10).

Several important questions about self-deception are under debate in the philosophical and psychological literature: Is self-deception intentional?

Does the self-deceived person actually hold contradictory beliefs? Does self-deception involve two selves or agents within the mind, a deceiver and a deceived? If self-deception and confabulation truly are related phenomena, and if the theory of confabulation in chapter 8 is on target, we ought to be able to learn something about self-deception by comparing it with theories of self-deception. The remainder of this chapter does this. We focus first on deception and lying and examine how they are different from confabulation; then we describe some philosophical theories of self-deception and compare them with the approach to confabulation developed in chapter 8.

9.2   Deception and Lying

Pam goes to a nightclub where she meets Sam. Sam is very tall, 6′10″ to be exact, and athletic-looking. To impress Pam, Sam tells her that he is a member of the Los Angeles Lakers basketball team. He explains that he has just made the team and will play a backup role this season. The following are important features of Sam's act of deceiving Pam into believing that he is a Laker:

1. It is false that Sam is a Laker   Is there a way around this? Suppose I deceive you into thinking that you have won the lottery. The drawing has not yet been held, so I don't know which numbers won, but I assume that yours did not, given the odds. But then it turns out that you actually have won the lottery. The best way to describe this seems to be that I *tried* to deceive you, but failed.

2. Sam believes that it is false that he is a Laker   Sam should have good reason for believing this. If the probability of its being true is too high, it becomes harder to see what Sam is doing as an act of deception (Schmitt, 1988). What if someone is self-deceived, though, and accidentally ensnares another person in his own self-deception? For example, consider the professor who is self-deceived about his status in the field and who deceives a student into also thinking that the professor is a great historian.

3. Sam engages in certain actions that represent him as being a Laker   This typically involves Sam claiming that he is a Laker. Such a claim would also be a lie, but Sam could effect his deception without saying a word, for instance, by wearing a full, official-looking Lakers warmup suit.

4. Sam intends to cause Pam to believe that he is a Laker by engaging in those actions   The way in which Sam causes Pam to believe that he is a Laker is important. If Sam uses some futuristic brain-stimulating device on Pam that causes her to believe he is a Laker, we do not regard him as having deceived her. Schmitt (1988) says that the deceiver must appeal to the *reasoning* of the deceived, by offering her justification. Notice that in order to offer Pam the right amount of justification, Sam must have some minimal capacity to represent her mind. In order for Sam to form the intention to cause Pam to believe that he is a Laker, Sam must also believe that Pam does not already believe that he is a Laker.

5. Pam forms the belief that Sam is a Laker Deception has a success condition involving the target of the deception that lying lacks. I can lie to you even if you fail to believe me. But I can only deceive you if I succeed in getting you to believe something I am aware is false. This criterion implies that Pam does not already have this belief. (Can I deceive you into *not* changing a belief you already held? When a wife believes her husband is faithful, but catches him in a compromising situation, he tries to explain it away to keep her from changing her belief.)

Deception and Mind Reading

Someone intending to deceive needs to sensitively gauge both what his victim believes and how much justification his victim has for those beliefs. We might distinguish between shallow deception, in which the deceiver uses few mind-reading capacities or none on his target, and deep deception, which is perpetrated with specific knowledge of the target person's mind. In shallow deception, the deceiver treats his victim the way he would treat any other victim. There are maxims of shallow deception: People have a natural inclination to believe positive claims about themselves, one should appear confident, and so on. One form of deep deception occurs when the deceiver takes into account differences between his own beliefs and those of his victim. The deceiver who assumes that his victim has basically the same beliefs as he does is still engaging in shallow deception. In shallow deception, the deceiver also does not model the mind of his intended victim. Deep deception should have a higher success rate than shallow deception, since it can warn the deceiver away from certain courses of action.

Many of the standard mind-reading tasks examined in chapter 5 have a strong epistemic character to them. When we think, "He believes it's in the chest, but it's really in the cupboard," we are tracking the person's beliefs and their causes. Flanagan points out that children's ability to lie develops in concert with their mind-reading abilities: "The competencies required for lying require a very elaborate theory of other minds. Children ... must understand that they themselves are being read for sincerity, reliability, and motivational state by their audience.... Saying what is false with the intention of producing false belief is rarely sufficient to produce false belief in a savvy audience" (1996, 107).

Another way to show the mind-reading component in full-blown deception is to observe what happens to our intuitions when we vary the amount of knowledge the victim has on the topic he is deceived about. Schmitt describes a case in which Sherlock Holmes is attempting to deceive

Dr. Watson into believing that p, where Watson has some evidence that not p: "When Watson possesses such evidence, Holmes must supply enough evidence for p to make the evidence Watson possesses sufficient to outweigh the evidence Watson possesses against p" (Schmitt 1988, 186). Holmes needs to know exactly how much evidence Watson does possess, and this is the sort of thing mind-reading processes are good for, especially those directed at beliefs, including any false beliefs. There are many ways the deceiver can offer his victim justification: his demeanor, the names he drops, the evidence he offers, etc. He can appear very confident, as in the case of confabulators and sociopaths. Being confident itself might fall under "offering a high degree of justification," in the case in which one's listeners will take confidence to indicate that the speaker is in possession of justification.

Lying

At least two mental states on the part of the liar are involved in any act of lying: (1) An intention to deceive by causing his victim to represent that x and (2) a belief (or assumption) that his victim does not already represent that x. The liar also believes (or perhaps merely hopes) that his victim will come to believe that x if he tells him that x in the right way. All of this requires that the liar possess the ability to represent the mind of his victim. As with deception, there is deep lying and shallow lying. Sociopaths seem to use shallow lying in that they tend to tell the same lies to everyone. They also generally believe that all people are gullible and easily lied to. Sociopaths do play to their victim's self-deceptions, but they typically do this with a generic list of self-deceptions. When they attempt to appeal to a self-deception their current victim does not possess, for instance by offering compliments about the appearance of someone who is not at all vain, it is often quite obvious to the intended victim what they are doing. If the sociopath does not represent his victim's mind, and we regard this as important for lying, this may explain why we sometimes are not certain whether it is right to say that the sociopath is lying (see section 4.4).

Creating a set of necessary and sufficient criteria for lying has proved difficult, but the following are some of the important characteristics behind the claim that Sam lies to Pam when he tells her he is a Laker:

1. It is false that Sam is a Laker If unknown to me you actually did win the lottery, and I tell you that you won in order to convince you to lend me money, I am not really a liar in this case. I intended to lie, but I did not succeed in lying. You might nevertheless regard me as a liar once you know what I did.

2. Sam believes that it is false that he is a Laker This may be stating it too strongly. Sometimes liars do not know the truth of the matter about which they speak. For instance, a man lies to his captors, saying he can lead them out of a wilderness area, so that they will not kill him. He does not know whether he can do this, but unlike the confabulator, the liar knows that he does not know.

3. Sam claims that he is a Laker to Pam   A liar engaged in deep deception will make this claim in just the right way that will cause his victim to believe him. Davidson notes a subcondition here: Sam has an "intention to be taken as making an assertion, for if *this* intention is not recognized, his utterance will not be taken as an assertion, and so must fail in its immediate intention" (1998, 3). The liar is defeated if his listeners think he is merely being ironic, or kidding.

4. Sam intends to cause Pam to believe that he is a Laker by claiming that he is a Laker   The fourth criterion implies that I believed that you did not already believe x when I lied to you. I need to know something about the nature of your beliefs because I need to offer you enough justification to change your existing beliefs.

Davidson argues that liars have two intentions: (1) The liar "must intend to represent himself as believing what he does not," and (2) the liar "must intend to keep this intention hidden" from his victim (1985, 88). The first criterion is implied by a combination of our criteria 2 and 3. Sam claims something he knows is false; this amounts to his representing himself as believing what he does not. Davidson's second criterion would be fulfilled by Sam as part of forming the intention described in criterion 4. Part of the intention to cause someone to believe something false is the intention to reveal certain things to them and to keep other things hidden. Again, a great deal of mind reading can go into first figuring out what the target knows and then figuring out how to manipulate that knowledge in an advantageous way.

Confabulators are not lying because they do not intend to deceive; they do not believe other than what they claim (or they believe what they claim); and also they are not modeling the minds of their listeners. The failure of confabulators to model the minds of their listeners is only an indication that they are not lying, as opposed to the other two criteria, which are closer to being necessary.

9.3   What Is Self-Deception?

The definitions developed by philosophers for self-deception contain a great deal of insight into this phenomenon. In this section I discuss the definitions proposed by Donald Davidson and Alfred Mele, and note how the characteristics they describe either apply to confabulators or do not. Then, conversely, I examine how the definition of confabulation given in chapter 8 relates to the phenomena and philosophical theories of self-deception.

Davidson defines self-deception, not in terms of the self-deceptive belief, but in terms of the feared thought. For example, Jan is self-deceived with respect to the proposition that she is an alcoholic.

Davidson's Definition

According to Davidson, an agent A is self-deceived with respect to a proposition q under the following conditions:

1. A has evidence on the basis of which he believes that q is more apt to be true than its negation (A is aware that the evidence is in favor of q.)   Recall that Davidson holds that all cases of self-deception involve tension, and in his view the evidence that A possesses seems to be the cause of his tension. Confabulators do have evidence in favor of, e.g., the thought that one is paralyzed, when the doctor tells them so, but they do not believe that this claim is more likely than its contradiction. One reason for this is that the evidence seems to slip out of their minds very quickly.

What about the case, though, of someone who genuinely does not know whether p or not p is true, but deceives himself into believing that it is? Sam tries out for the Lakers and seems to do fairly well, based on his impression, but also on what the coaches and his fellow players say. At the end of the tryout, he judges that he has a 50–50 chance of making the team. However, over the next few days, as he waits to hear, he deceives himself into believing that he has made the team (another person with a pessimistic bent might start to believe that he did *not* make the team). There may or may not be great tension here.

2. The thought that q, or the thought that he ought rationally to believe that q, motivates A to act in such a way as to cause himself to believe in the negation of q In cases of confabulation in the clinic, it is the examining doctor who raises the thought that q (e.g., your left side is paralyzed). Then in some cases the patient's cognitive system produces the opposite belief, which can exist unchecked by anything that might normally downgrade it to a mere wish. Beliefs that are formed solely because their contradictories are frightening are ill-grounded beliefs. The fear that q is an illegitimate top-down influence involved in the formation of the belief that not-q, we might say. The confabulation literature indicates that this process can go on unconsciously and unintentionally. Hence it is possible that in confabulating patients, the thought they are paralyzed, blind, amnesic, etc. does play a causal role in generating the confabulation. For an Anton's patient, the thought that he is blind, as suggested by the doctor, may play a causal role in getting the person to say, "No, it's just dark in here." Nevertheless, confabulators are not aware of this process and tend to believe the confabulation.

3. The action involved may be no more than an intentional directing of attention away from the evidence in favor of q; or it may involve the active search for evidence against q One good feature of this criterion is that it also applies to cases of tension-free self-deception. People who are self-deceived about how good they are about something may only review the positive feedback they got about their performance and forget any negative feedback. The anosognosic does not need to direct his attention away, since attention is gone from the crucial domain. Notice also that directing the attention away is a technique for preventing clear representations from being formed. This point can be made more general: Self-deception is facilitated when representations are degraded *by any means*. The mention of searching for evidence also highlights the curious way that confabulators do very little of this, something that may be related to their depressed autonomic activity. They are perfectly happy with the state of their beliefs.

Mele's Definition

Perhaps the most-discussed recent approach to self-deception comes from Alfred Mele (1987, 2001). He argues that the following criteria are jointly sufficient for S to be self-deceived in acquiring the belief that p (2001):

1. The belief that p which S acquires is false   One problem with such a criterion is that we sometimes seem to know that someone is self-deceived without knowing that what he believes is false. This can happen with self-deception about the future, in cases where we do not know whether the crucial belief is false, but we strongly suspect it is. What about the case of someone who deceives himself into believing that he will win the lottery? He believes it so strongly he gives away his car, etc. Suppose then that he really does win the lottery. Does that mean that he did not in deceive himself the first place? The problem is that he must have arrived at the belief in a way that produces ill-grounded beliefs. In the same way, one can imagine a confabulator making a true claim out of luck.

   Mele (2001) holds that self-deception inherits a falsity condition from deception; that a person is *deceived* in believing that p only if p is actually false. His approach here would be to allow that the lottery player might be self-deceived, but that we cannot say he was self-deceived *in believing he would win*. My intuitions pull in both directions here, but the emphasis on falsity may be because we have so little knowledge about the exact nature of the epistemic problems in self-deception. I argued in chapter 8 that ill-groundedness was more important than falsity in the case of confabulation, primarily because our efforts need to go into determining the exact nature of the ill-groundedness rather than the truth of the claims. If Davidson is right, self-deception is also an epistemic phenomenon, so something of the same sort may happen with it. Once we understand the phenomenon of self-deception, we will know what kind of process it is, and we then need to rely less on the question of which particular beliefs are true or false.

2. *S* treats data relevant, or at least seemingly relevant, to the truth value of *p* in a motivationally biased way This criterion applies both to tension-filled and tension-free cases of self-deception. People who fancy they are far above average often recount and savor any remark by others or event that they feel provides evidence for this. At the same time, they discount information that might challenge their happy beliefs. Confabulators often seem to be treating disconfirming evidence in a motivationally biased way, but any motivations are apparently far from consciousness.

3. This biased treatment is a nondeviant cause of *S*'s acquiring the belief that *p* Beliefs formed in this way are presumably ill-grounded. Bias in self-deception and brain damage in confabulation may have similar effects though: Wishful thoughts are not stopped from becoming beliefs. Mele (2001) fleshes out his third criterion by describing several different ways in which a motivational bias might cause someone to believe that *p*. Some of these also work as maintenance processes for self-deceptive beliefs that have already been formed:

a.   Selective attention to evidence that we possess

b.   Selective means of gathering evidence

c.   Negative misinterpretation
   We fail to count a given piece of evidence, *e*, against *p*.
   We would count *e* as evidence against *p* if we did not have a motivational bias against believing that *p*.

d.   Positive misinterpretation
   We count a piece of evidence, *e*, as being evidence for *p*.
   If we were not motivationally biased, we would count *e* as evidence against *p*.

One interesting technique for selective gathering of evidence is gaze aversion. We do not look directly at the thing that makes us feel bad. For example, I crumple the front fender of my car in an accident, and later I deliberately do not look at it as I pass the car on my way out of the garage. The neglect patient with anosognosia does not need to direct his attention away from his paralysis or visual loss; his neglect is enough to do this. Some of the behaviors of anosognosic patients may also fit the description of negative misinterpretation. The patient who denies paralysis fails to count the fact that his left arm did not move as evidence supporting the doctor's suggestion that it is paralyzed. Instead, he concludes that it did not move because he is tired, sore from arthritis, and so on.

4. The body of data possessed by *S* at the time provides greater warrant for not-*p* than for *p* Depending on what it means to "possess" data, this criterion may imply that *S* should know that her belief is ill-grounded. The problem in confabulation is that it is not clear that the patient really *pos-*

sesses the data that contradict the confabulation. The needed representations might be degraded or inaccessible or may fade from his mind before he appreciates what they mean.

Confabulation and Self-Deception

Next let us look at the criteria for confabulation developed in the chapter 8 and assess how they apply to self-deceived normal people.

1. Jan claims that *p* A self-deceived person might claim his belief outright, although that tends not to happen. Perhaps at some level the self-deceived person is afraid that if he states his belief explicitly to another person, that person will graphically describe just how ill-grounded it is. This holds primarily in tension-filled cases. People in a state of tension-free self-deception seem to be more likely to voice their self-deceptive beliefs, and certain types of self-deceived people may brag at length about the very topic on which they are self-deceived. There are also those interesting occasions when someone in a state of tension speaks at length about the topic of his self-deception, apparently in an attempt to convince both himself and his listeners.

2. Jan believes that *p* As with the confabulator, the self-deceived person typically holds the relevant beliefs, as evidenced by their propensity to act on them. One feature the two share is that both confabulators and self-deceived people may have islands of ill-grounded beliefs amidst a basically well-grounded set of beliefs.

3. Jan's thought that *p* is ill-grounded One way a self-deceptive belief can be ill-grounded is if a feared thought played a role in its genesis. Coming to believe that *p* merely because you fear that not-*p* is not a good epistemic strategy, although there may be nonepistemic reasons for it, such as the idea that denial protects us from becoming depressed. As Davidson notes, the primary reason for the ill-groundedness or "weakness of the warrant" in the case of self-deception is that the person is aware at some level of evidence against his belief.

4. Jan does not know that her thought is ill-grounded Many tension-free self-deceived people do not know that their self-deceptive beliefs are ill-grounded. There is another class of self-deceived people who at some level suspect their self-deceptive beliefs are ill-grounded. For the self-deceived person, not knowing that a belief is ill-grounded involves a battle between belief-protecting processes and belief-dislodging processes. The person does not let himself realize, or come to know, that *p* is ill-grounded. This battle cannot take place in the mind of the confabulator because the belief-dislodging processes are malfunctioning.

5. Jan should know that her thought is ill-grounded This still seems to hold in the case of the self-deceived. Neither the confabulator's nor the self-deceived person's cognitive system is functioning optimally. The confabulator has brain damage, while the self-deceived person's mind is in a suboptimal operating state from an epistemic point of view. We hold the self-deceived person more at fault than we do the confabulator, however, because there is reason to think the self-deceived person *can* use the processes needed to dislodge his belief. What makes us call someone self-deceived seems more to be that she does not know something she *should* know. Self-deceived people can be difficult to relate to or work with, since there is always the chance that their self-deception, or their self-deceptive tendencies will interfere in interactions. If we know that a person is self-deceived about how good he is at something, for instance, we also know he will try to take any credit or other rewards for being good at this activity, possibly *from us*. This criterion applies to both tension-free cases of self-deception and to those with tension. Notice, though, that the sense of "should" employed here has shifted from the more functional sense applied to confabulators to that having more to do with responsibility and ethics because we see the problem as being in the person's power to correct.

6. Jan is confident that *p*   This criterion applies to tension-free self-deceived people, but not to those with tension. Tension-free self-deceived people can be quite confident in their beliefs, and they may genuinely not be in possession of any conflicting information. Self-deceived people with tension are not confident, however, although they might defensively claim that they are.

According to this discussion then, when a self-deceived person who experiences no tension asserts a self-deceptive belief, he is confabulating. The academic who is far worse than average but believes he is better than average is confabulating if he asserts that he is better than average. The tone-deaf singer who says, "I'm good enough to be on TV," is confabulating. The two phases of confabulation are there: First, an ill-grounded belief is created, and second, the person lacks the ability to properly check thoughts of that sort and prevent himself from making claims based on them. These people should know that their beliefs are ill-grounded (criterion 5). This can be seen in the way we hold them responsible for not knowing.

9.4   The Maintenance of Self-Deceptive Beliefs

Talland gives the following description, of a Korsakoff's patient: "A brief moment of insight into her condition produced no words, only a visible emotional shock, and tears of distress and bewilderment. Confabulation was resumed a minute or two later, by which time the cognitive as well as the affective registration of her disability had, by all appearances, vanished"

(1961, 369). Confabulators suffer from doxastic weakness; they only hold on to the contradicting thought for a while; as a belief, it seems to lack staying power. Though they may assent to it, they never quite solidify the admission of illness into an enduring belief. Self-deceived people may also admit their deceptions briefly, then sink back into them. This indicates that there are two parts to self-deception: creating the deceptive belief, then failing to discard it in a lasting way.

It is easy to maintain self-deceptive beliefs when there is no tension, but maintenance becomes more complicated when tension is present. Davidson contends that a feared belief can not only cause a self-deceptive belief to come into existence but also plays a role in maintaining it: "The state that motivates self-deception and the state it produces coexist; in the strongest case, the belief that p not only causes a belief in the negation of p, but also sustains it" (Davidson 1986, 89). There needs to be a sustaining process in the normal mind for self-deception to work. Since the frontal checking processes are able to operate, some other force in the mind needs to keep them away from the self-deceptive belief.

Normal people can sometimes enter into states of brief, intense self-deception, that cannot be sustained. If you have ever received shocking news over the telephone, for instance, news about the death of a young friend in an accident, you have likely seen the initial self-deception process at work. One's first response is denial: "No. This is *not* true; it's a sick joke, or some sort of mistake." But the denial cannot be sustained, and within several minutes' time, we come to accept the truth of what the voice on the phone says.

Another way to maintain a self-deceptive belief is to use brain processes that function specifically to suppress other states and processes. Recall Schnider's (2001) theory from chapter 3. He and his colleagues hypothesize that a process in the orbitofrontal lobes is responsible for suppressing irrelevant memories when a person is trying to recall an autobiographical event. If Schnider is correct, this means that the brain contains processes capable of suppressing memories. Fingarette argues similarly that self-deception involves the brain's use of functions devoted to focusing and maintaining attention: "The aura of paradox has its real source in the failure to appreciate properly how our mind ordinarily, normally and familiarly works in every waking moment" (1998, 289–290).

One problem in some cases of self-deception is the question of how we are able to repress anything nonconsciously, but Fingarette points out that this happens all the time: "The crux of the matter … is that we can take account of something without necessarily focusing our attention on it. That is, we can recognize it, and respond to it, without directing our attention to what we are doing, and our response can be intelligently adaptive rather than merely a reflex or habit automatism" (1998, 291). We see this happen every day according to Fingarette: "Just as I can avoid focusing my attention on the potentially distracting sounds of the passing cars, and do so without focusing my attention on the fact that I am doing this, so too I can take account of something potentially emotionally traumatic, and for that reason avoid turning my attention to it, and do all this without turning my attention to the fact that I am doing it" (1998, 295). These normal attention-maintaining processes may be used for self-deception.

In chapters 3 and 4, we saw that damage to the mediodorsal nucleus of the thalamus is often present in confabulating patients. Anatomists Ray and Price note that the thalamic mediodorsal nucleus may function to block cortical activity: "If thalamocortical activity provides a mechanism for sustaining activity related to concentration on a particular task, inhibition of [the mediodorsal nucleus] may block sustained activity, in order to switch between different patterns of thalamocortical activity, or to suppress unwanted thoughts or emotions" (1993, 29).

There is also an emotional component involved in sustaining self-deceptive beliefs and beliefs that give rise to confabulations. We saw the importance of emotional responses in decision making when we examined patients with orbitofrontal damage in chapter 4. The failure of their brains to generate the proper inhibitory emotions seemed to make these patients behave in disinhibited ways. Emotions may also play a role in getting self-deceived people to give up a self-deceptive belief permanently. Without the proper emotional response, the person might admit his evidence is weak, but then shortly thereafter reaffirm the self-deceptive belief, just as confabulators do. In many cases, the self-deceived person avoids thinking a crucial thought with an emotional feeling of conviction, in the way that sociopaths do not feel emotions deeply. Sociopaths who are capable of thinking a crucial thought (I should not be hurting people like this) with conviction are self-deceived; those who are not capable of such conviction are self-deceived in a tension-free way. Even in normal people, to be self-deceived is not to *never* think a feared thought. No one can avoid thinking such a thought at least sometimes, in the same way that one cannot avoid thinking of camels once ordered not to. But one can think a thought without conviction, especially if one's feelings of conviction are weak or disorganized.

9.5   Questions about Self-Deception

Does Self-Deception Involve Holding Contradictory Beliefs?

The quick answer to this is "no," because a self-deceived person may be able to prevent the conflicting belief or representation from forming. There are also other ways that the self-deceived person can experience conflict without possessing two contradictory beliefs. Mele (2001) considers cases presented by experimenters that are claimed to involve two contradictory beliefs. Subjects who wrongly state that a tape-recorded voice is not theirs nevertheless show SCRs that are consistent with voice recognition. According to Sackheim and Gur (1985), these responses indicate that the person also *does* believe that the voice on the tape is his own. "It is unclear," says Mele, "that the physiological responses are demonstrative of belief" (2001, 83). This seems correct, but there may be some cases where the presence of an SCR may indicate conflict or tension, as in the case of the SCR sometimes produced by liars.

A person can also represent something as being the case without having a belief that it is the case. In addition to beliefs, the brain also contains topographic representations, which are a variety of analog representation. One source of difficulty may be that we have trouble thinking of the information contained in the brain's topographic, or analog representations as consisting of beliefs (Hirstein, 2000). Searle gives an example of a problematic epistemic state, which we are not sure is a belief or not: Suppose you were to enter your office one day and find a huge chasm in the floor. You would be surprised, Does this mean that you believe that your office floor is solid, or that you have a belief that there is no chasm in the middle of your office floor? We have a robust intuition that there is something wrong in ascribing such beliefs to you. I suggest that this is because, while you represent your office floor as solid, you represent it in an analog or topographic form, rather than in the form of an explicit belief. You have been in your office hundreds of times, and your right hemisphere body-in-space representation system has represented the room each time. An analog representation can be in conflict with a belief in the same way that a picture and a sentence can be in conflict.

This suggests a way to resolve a question about self-deception. How can a person selectively avoid evidence for the belief that *p* without in some way believing that *p*? How else does he know what evidence to avoid? One way to deal with this is as follows. The information that *p* is already represented in analog form. There is conflict in the person's mind, but the conflicting information is represented in two different forms, conceptual and analog. This is different from holding two contradictory beliefs in full conceptual form. What may be happening is that the brain has a way of preventing certain types of analog information from being represented in conceptual form, from being explicitly thought and believed.

Is Self-Deception Intentional?

Looking at the question of whether our concept of deception allows nonintentional forms of deception turns out not to be too helpful because our uses of "deception" in this way are extremely broad. Our uses of "deceive" are similar to our uses of "distract" when it comes to the intentional and unintentional distinction. I might intentionally distract you when I cough during your backswing in golf, or I might do the same thing unintentionally. We speak of deception in the wild, for instance, in cases where there is nothing like an intention. Gianetti (1997) mentions the orchids that attract male bees by producing a scent similar to that of their females. Are the orchids *deceiving* the bees? If this is deception, such cases tend to work against the notion that self-deception is intentional, in that they provide a clear example of nonintentional deception. Perhaps the reason we say the animals are deceiving each other is that the closest human equivalent (e.g. dressing in disguise) is a genuine case of intentional deception.

Treating this question properly would involve tackling the difficult question of what it means for an act to be intentional. This is a huge area that we need not get into here, but we can point out that one reason sometimes given for the claim that self-deception is intentional seems to be obviated by information gained from confabulation. Self-deception is selective in that self-deceived people select a special set of beliefs that they protect. Bermudez (2000) says that the selectivity in self-deception is best explained by positing that the person intentionally brings it about that he believes that *p*. There is thus an intentional selection process at work. However, as we saw in chapter 6, patients with anosognosia will also selectively deny one problem while admitting others. It is widely agreed that this denial is not a case of intentional deception—either of the patient himself or of the questioner. The selectivity in confabulation is caused by the nature of the damage, not by the patient's intentions. There is thus a clear empirical example of how selectivity could exist without the claimed intentions.

The problem with selectivity is that one can know that it is there without knowing anything at all about what is doing the selecting, and how. Several processes below the level of full intentional action can create this selectivity. One can imagine a simple process that keeps items from coming into consciousness on the basis of a nonconscious reaction to that information. We saw in earlier chapters that the brain can respond to stimuli even though the person is not aware of that response. These unconscious states need to distinguish between positive and negative reactions. When the reward value associated with a thought or stimulus is negative enough, that thought can be prevented from becoming a conscious thought.

There may be a connection between the amount of tension present and whether or not the deception is intentional. The clearer the person's awareness that he possesses contrary information, the stronger, i.e., the more intentional, the deception needs to be. Self-deception only needs to be intentional if the feared thought comes fully into consciousness. Intentional self-deception and nonintentional self-deception cause a content-specific abeyance of the checking processes, but by different causal routes. A person can stop explicit checking from happening, but at the same time nonconscious cycles successfully complete, causing anxiety (including autonomic activity) and a sense of tension in some cases. Notice, though, when tension is absent, that the more intentional the self-deception is, the less likely we are to complain that the person involved is self-deceived. The man who deliberately sets his clock ahead fifteen minutes because it tends to get him to work on time might be said to be deliberately deceiving himself, but we do not speak of such people as self-deceived.

Degraded Representations

What is the point of saying, "Don't tell me the details, I don't want to know," when someone says that a mutual friend has died in a horrible accident? One may even know exactly what the person will say. The problem is that this will make the representation more vivid, and it will be responded to more strongly. There seems to be a version of the degraded representation principle given in chapter 8 that applies to self-deception also: If a person has a diminished capacity to represent events of type x, that person will be more likely to self-deceive about events of type x.

Apparently degradation can come about by intentional or unintentional means. In confabulating patients, it occurs as a result of brain damage. In normal people, as in the example, it seems to be brought about intentionally in some cases. One way to see how degraded representations can result in self-deception is to look at cases in which people are deceived about how good they are at a certain activity because they do not have clear, detailed representations about what it is like to be good at something. Imagine a mediocre golfer who is laboring under the illusion that he is playing at the level of professionals. He does not know enough about the mechanics of the golf swing to see the huge differences between the fluid, powerful motions of the pros he sees on television and his swing. He keeps score creatively, forgetting to assess penalty strokes and allowing himself to replay muffed shots. He also seems not to be aware of how much professional golfers practice. If he was aware of these things, one suspects it would be sufficient to cure him of his self-deception.

Desire

The question of whether a certain wishfulness should be an essential feature of the concept occurs in discussions of both confabulation and self-deception. In confabulation, this wishfulness is clearest in anosognosia. The hemineglect patient and the Anton's syndrome patient are just fine, according to them. However, there are other confabulators who make claims that have negative implications for them—the Capgras' patient who says his wife is an impostor, for instance. Mele describes a phenomenon he refers to as "twisted self-deception" which seems to show that the requirement that the person desire that p is not necessary. He gives the following example: "An insecure, jealous husband may believe that his wife is having an affair despite his possessing only relatively flimsy evidence for that proposition and despite his not wanting it to be the case that she is so engaged" (Mele, 2001, 4).

Confabulators often do seem to create confabulations that conveniently satisfy desires, albeit in an indirect manner; for instance, the patient's desire that he be of sound body after a right-hemisphere stroke. Other confabulations are simply ill-grounded claims that are not the expression of any particular desire. The distinction between normal and twisted self-deception can also be assimilated into the theory outlined in chapter 8 (see section 8.3). The difference is between the creative stream's being wishful, or paranoid. In either case, frontal checking processes might not be able to sufficiently counter the creative stream. The problem in both confabulation and self-deception is the failure of frontal processes to discard ill-grounded beliefs. This is what is crucial, not the particular valence of the products of the creative process.

Self-Deception and Weakness of the Will

At the end of the process of cognition, when the time comes for the conclusion of a piece of thinking to give rise to intentions and actions, the human cognitive apparatus is prone to a number of problems, including one known as *akrasia*, or weakness of the will. Imagine the following scenario (from Donald Davidson): You just tucked yourself into bed on a chilly night. As you slowly begin to drift off to sleep in the peaceful warmth under the covers, up pops the realization that you have forgotten to brush your teeth. You lie there torn right down the middle between the belief that you must brush your teeth every night, and the slothful desire to remain where you are. As with inhibition, perhaps only some sort of emotional reaction can get us up out of bed, and without this, all the good reasons in the world cannot stir us.

Weakness of the will often involves the failure to stop doing something a person is currently doing—smoking, drinking, etc.—in which case it could be viewed as a type of inhibition failure. We might consider the behavioral disinhibition of the frontal patients discussed in chapter 4 as a severe form of weakness of the will. If we understand the making of a claim as a kind of action, confabulation falls under weakness of the will, just as it counts as a type of disinhibited behavior. Since claiming is an action, we might see certain types of claiming of self-deceptive beliefs as involving weakness of the will also. Bragging, for instance, often involves claiming self-deceptive beliefs (but it need not; the "brag" might be true). So the outcome of weakness of the will is an intention (e.g., to smoke), and likewise the outcome of self-deception can be an intention to claim that p.

9.6 Self-Deception and Mind Reading

If Sam is engaging in deep deception when he deceives Pam, he is employing mind-reading capacities to understand Pam's mind. Hence his mind contains a representation of her mind. He concocts his plan for deception using this model of her mind to increase the probability of the deception working. When deep deception occurs, the deceiver has two separate processes at work in his mind: a process that creates the deceptive claims and a process that models the mind of his victim. In cases of interpersonal deception, the mind-simulating system is treated as a representation of the mind of the target person. I suggested in chapter 5 that on other occasions the mind-reading system functions, not in this role, but as an inhibiting and checking system. This reveals an interesting connection between self-deception and other-deception. In both cases, one is making claims designed to elude the checking processes that constitute the inhibitory system. It also explains why chronic deceivers also tend to be self-deceivers. The same processes are needed to prevent both types of deception. This is why sociopaths lie so effortlessly. They simply make their normally self-deceptive thoughts public, and others with similarly weak checking processes will believe them.

Many people can get carried away in one person's public self-deception. This only makes sense—what works on one person may well work on many. Sociopaths often seem to be self-deceived, but as with the confabulator, there does not seem to be much tension inside them. It may also be that sociopaths have specific techniques for quieting down, diverting, and confusing the checking processes.

The checking processes can embody a generic set of reactions: How people in general would react if you acted in a certain way. We often speak of people as self-deceived when they convince themselves that the majority of people are wrong about something; that the received, or orthodox views are mistaken. Sometimes the issue in question is an ethical one. If self-deceived people suppress this inner audience, confabulators lack it entirely. The existence of the mind-reading system may also help explain why some theorists of self-deception posited a deceived second self within the person. Since the mind-reading system functions as a simulation of the mind of the target person, it has selflike characteristics. Trying to "fool" this system becomes very like trying to fool another person.

While confabulators are not themselves attempting to deceive, they would be good targets for someone who was attempting to deceive.

Confabulators lack the critical faculties to apply to their own representations and to those created by others. Confabulators may also have difficulty detecting confabulation in other people. In an informal test on a split-brain patient (L.B.), we read several examples of confabulation to him. He was unable to notice anything wrong. When we said such things as "What if the person is merely making up a story and has no idea whether it is true or not," his response typically was a bland, "Well, if he was doing that, that was wrong." There are further questions we might go on to ask to probe the links between self-deception and mind reading broached here: Are self-deceived people generally deceived about what others think of them? Do self-deceived people also deceive themselves about what is happening to others? What does a self-deceived person do when he sees someone in his situation?

9.7 The Neuroscience of Self-Deception

If we understand obsessive-compulsive disorder correctly, it is the opposite of self-deception. One suspects that people with OCD have difficulty deceiving themselves, although this may vary with the particular knowledge domain. Whereas the self-deceived person might say to himself, "It'll be OK if I don't brush my teeth tonight," the person with OCD will get up and brush his teeth again and again. The nagging thought, so successfully suppressed by self-deceivers, rages out of control in OCD, cannot be ignored, and must be acted on. Some of us self-deceive too much, others of us not enough. OCD sufferers need some way to make the obsessive thought go away and stop interrupting their mental lives. The person who is self-deceived about symptoms that may indicate a health problem is sliding closer to the anosognosic, while the person with hypochondria is closer to the person with OCD.

Gazzaniga provides some interesting connections between his concept of the left-brain interpreter, lying, and self-deception: "What is so adaptive about having what amounts to a spin doctor in the left brain? Isn't telling the truth always the best? In fact, most of us are lousy liars. We become anxious, guilt ridden, and sweaty. As Lillian Hellman observed, guilt is a good thing, it keeps us on the straight and narrow. Still, the interpreter is working on a different level. It is really trying to keep our personal story together. To do that, we have to learn to lie to ourselves" (1998, 26). Rather than working "on a different level" as Gazzaniga proposes, I argued in chapter 4 that these autonomic signs of lying are connected with orbitofrontally based processes that oppose the glib, smoothing function of left-hemisphere explanation-producing processes. Those processes are capable of casting doubt on the explanations of the interpreter, and preventing them from becoming beliefs.

Besides the comparison with OCD, there are other ways contemporary neuroscience research can shed light on self-deception. One recent study shows that deliberately not thinking about something can have the desired effect of making it more difficult for that information to get into consciousness later on. Anderson and Green performed an interesting experiment in which they asked normal subjects first to remember the second of forty pairs of unrelated words. Once they had learned the pairs, in the second phase of the experiment the subjects were shown the first member of a word pair, together with another cue that instructed the subject to either think about the second member of the pair or not think about it. "For the latter pairs, we emphasized that subjects should not allow the associated memory to enter consciousness at all" (Anderson and Green, 2001, 366), although one wonders how successful they were at this task. However, the subjects did in fact show poorer recall later of words they were instructed not to think about.

Electroencephalograph (EEG) expert Eric Halgren describes a possible mechanism of unconscious self-deception. High-resolution EEG appears to show that the emotional evaluation of a stimulus is independent of the cognitive evaluation of it (at least in early stages), and completes sooner: "The fact that the limbic system is heavily activated beginning with the early stages of event encoding implies that emotional information is integrated in very early stages of event encoding. The prominent limbic activity beginning 120 ms [.120 seconds] after stimulus onset permits limbic input to shape the content of the encoded experience rather than simply react to its content" (Halgren and Marinkovic 1995, 1146). These authors do not shy away from the remarkable implications of this: "Conceivably, this would allow the limbic system to contribute to the myriad psychological defense mechanisms (repression, denial, undoing, projection, displacement, etc.) that may distort or eliminate the conscious experience of an emotionally significant event" (Halgren and Marinkovic 1995, 1146). One must be careful with a concept like emotional information, however, especially in cases such as this where any emotions are unconscious. There is presumably a set of cortical areas and subcortical organs that are directly responsible for embodying conscious emotional states. There are also a wide variety of processes and organs that support this activity, including the entire limbic system. The question of what in this larger set should be labeled emotional information, emotional processes, etc. is a difficult one, but as long as we can distinguish among those areas and processes directly involved in embodying or realizing conscious emotional states, and the many supporting processes, this may be all we need to avoid uncertainty about what is meant by the uses of the term "emotion" in these contexts.

Self-Deception and Reasoning about Probability

There may be a connection between the failures of estimation seen in confabulating patients (see chapter 3), and the failure of many self-deceived people to sum the different sources of evidence against their self-deceptive beliefs. The man whose wife is having an affair fails to add the different sources of evidence, and tries to explain each one away separately. The first piece of evidence should indicate a certain probability that the feared belief is true. When the second piece of evidence appears, it is often wise to set the resulting probability higher than merely adding the first and second probabilities taken independently. A third piece of evidence is then often enough to push us to confirmation, to acceptance of the belief. People who in general are not good at assessing probabilistic evidence may be more prone to self-deception. This self-deception is less intentional than that of someone who can correctly sum probabilities, but who disables these processes to prevent them from coming up with the feared result—the awareness that the probability of the feared thought is high.

It is easy to produce some sort of evidence that conflicts with a feared thought, and without a sense of probability that evidence can appear to neutralize the feared thought. The difficulty with estimation that confabulatory patients have may also appear as difficulty determining the stronger of two opposing claims in the cognitive realm. One often encounters people who consider any dispute to be a draw, as long as each side can offer some plausible reason in response to objections from the other. A frequent technique in self-deception is the use of these "balancing" thoughts. Any reason, however weak, that goes against the strong arguments in favor of the feared thought can push it out of consciousness. These balancing thoughts may work because the self-deceived person is able to inhibit the processes we employ to assess probability, or lacks them to begin with. This type of deficit is often part of the mechanics of self-deception. Suspicions that, e.g., one's wife is having an affair can be kept at bay by any sort of disconfirming evidence, no matter how weak.

9.8   Conclusion

Confabulation and self-deception are different but related phenomena. They both are due to a failure of the brain's belief checking and improving processes. However, the question is why we are prone to self-deception and confabulation in the first place. Would it not be far more effective from an evolutionary perspective for us to be completely honest with ourselves? Why must we try to hide things from ourselves? What does the existence of these phenomena tell us about who we are and how we evolved? Do confabulation and self-deception have good or healthy functions? For instance, one healthy function of confabulation may be protection. It assists brain-damaged patients by delaying the full realization of their disability until they are ready to cope with it. A second possible healthy function is cognitive. Rather than reorganizing one's entire belief system when it is found to be flawed (for instance, one's belief that one's body is sound, or that one's wife is faithful), sometimes it is better to cover up the problem, at least until a large enough set of problems develops to force a wholesale restructuring.

If you look at your average contemporary person, the potential for tragedy is immense. The people and things we love and value are strewn across the globe. Any number of health disasters can befall you or them. The average drive to work carries with it the possibility of a serious car accident. In general, the truth of things is painful, and realizing the truth can be depressing. Certain forms of anxiety attack may occur when the self-deceptive processes break down and consciousness is flooded with the truth of things. The truth is depressing. We are going to die, most likely after illness; all our friends will likewise die; we are tiny insignificant dots on a tiny planet. Perhaps with the advent of broad intelligence and foresight comes the need for confabulation and self-deception to keep depression and its consequent lethargy at bay. There needs to be a basic denial of our finitude and insignificance in the larger scene. It takes a certain amount of chutzpah just to get out of bed in the morning.

10 Epilogue: Our Nature

To fear death, gentlemen, is no other than to think oneself wise when one is not, to think one knows what one does not know.
—Socrates (399 B.C.)

10.1 The Meaning of Confabulation

The creative process of confabulation, its character and persistence, is emblematic of what it is to be human. Gazzaniga's name for it, "the interpreter," is not inappropriate; it places actions and events in a particularly human context we find easy to understand. However, this is exactly why we need to be skeptical about the products of this process. It is capable of providing the peace of mind an explanation brings, whether or not that explanation is a good one. The fact that we find stories so easy to understand and convenient to use is no doubt related to the ease with which we are tricked by them. In the healthy human mind, this interpreting, creative faculty does not work alone; it works in tandem with processes designed to evaluate its interpretations, and if necessary, prevent them from being expressed or acted upon.

This interaction—the dynamic equilibrium between the creative and inhibiting forces—greatly increases the power of human thought. It contains everything necessary to carry out a schema consisting of hypothesize–check hypothesis–modify hypothesis, such as that followed in the scientific method. We can model actions and employ frontally based checking processes to determine our best course of action. The structured interchange, or dialectic, between creation and inhibition is often open to view within our mental lives. We are often consciously aware of the products of the creative processes as well as the inhibitory work of the checking processes. The strength and character of these two process types within each of us is also an important part of what we know as personality. Recall the idea that recurred throughout our survey of the confabulation syndromes—that the patient's prior personality is one of the factors that determine whether a certain brain lesion will cause him or her to confabulate. The inhibitory processes are also an important part of who we are, both as individuals and as a species. We are not only what we do, but to a great degree we are also what we do not do and do not say, as the case of Phineas Gage shows. People vary in how many individual acts of inhibition occur in them during an average day. Some people are guarded in what they say, while others seemingly cannot think something without saying it. The inhibitory power of the checking processes also helps to determine whether we are seen as confident or unsure by our listeners. And the productive power of the creative processes can determine the degree to which our communications are creative.

Confabulation might be counted among the human foibles, the set of fallacies and cognitive errors to which our thinking processes are naturally prone. Every physical process has its limitations. Each perceptual organ, for instance, is sensitive to certain properties in the environment, but blind to a huge set of others. Blind spots can be dangerous though. How much damage is done by people who don't know that they don't know enough about an issue to make a decision? The problem is not so much ignorance as it is complacent ignorance; people who don't know that their knowledge is insufficient to act on are dangerous. Memories of crucial mistakes must be marked in some way to prevent their reoccurrence. But one should not dwell too much on this; obsessive-compulsive disorder looms at the other extreme, threatening to turn all thought into pointless rumination.

The remarkable power of human representation may also be an indirect cause of confabulation. Representational power gives us the ability to formulate huge numbers of thoughts that we do not know the truth value of. One way to construct an organism that will maximize its epistemic potential is to make the realization of ignorance a negative, unpleasant state for that organism. Too often, however, lack of knowledge has a paralyzing effect on us. Conversely, finding the solution to a problem produces a positive state. Perhaps, then, confabulation serves to allay fear of the unknown. But why design an organism that is afraid of the unknown, and not, say, excited or fascinated? There may be danger in the unknown, but there may also be items of value. This fear comes to us all, especially when the stakes are high. The scientist is happy when he or she comes up with a hypothesis to explain an unexpected result, but then becomes edgy and nervous when new evidence does not support the hypothesis.

Why are we drawn to the confident, confabulatory person, while the doubt-filled person is always a bit unpleasant to be with? Confident, confabulatory people can be quite convincing until we detect the falsity of their statements. Partly because of the natural emotion-sharing processes among humans, the unsure person makes us uneasy. Whatever the reason, epistemic overconfidence can be a successful strategy in social interactions, and it may be that confabulation and self-deception derive from this in some way. Perhaps the answer is that we are programmed to be overconfident for social reasons. Being overconfident is not good from the standpoint of truth, but this may show that truth is secondary for us. Ideally, we would need to be overconfident socially, then more epistemically prudent in nonsocial realms, where this strategy would have much less use.

A final reason for the existence of confabulation may have to do with creativity. In the clinic, I am often amazed at how creative confabulations are. In confabulation, we may be seeing the human gift for creativity in a naked form. We are seeing the creations in the absence of other processes that assess their plausibility or practicality. Thus conceived, confabulation is a by-product of what is normally a potent epistemic strategy: over-generate hypotheses and filter them with checks (similar to Dennett's [1991] multiple-drafts theory). Our brains developed in such a manner that this type of filtering came into existence. I have cited a Jacksonian explanation for this. Recall the principle of brain construction from chapter 2, that in the brain's deeper levels are still-functioning ancient brain structures, which we share with many other species. Newer levels leave older levels intact, but are capable of filtering their representations.

10.2   Further Questions

Our goal has been first to assemble the data on confabulation and attempt to understand what it is. If it is based on epistemic ill-groundedness, one obvious next step is to assess each of the knowledge domains and its accompanying sets of processes to determine the exact nature of the ill-groundedness. There are several other interesting issues that arise at this point and they are discussed in the following sections.

Confabulation and Consciousness

One way to approach the question of how confabulation relates to consciousness is to ask whether there is evidence that the portions of the orbitofrontal lobes we are interested in are capable of sustaining conscious states or parts of conscious states. We have seen that several writers posit that they are. We have also seen that the phenomena of confabulation have led some theorists to make inferences about the nature of consciousness. Some of these writers believe that consciousness is an important, effective brain process (e.g., Schacter 1990; Kinsbourne 1988), while others hold a deflationary view of it (Dennett 1991; Gazzaniga 1998; Wegner 2002). One might approach this problem by combining the information here about brain damage in confabulation with what is known about the parts of the brain that are crucial for consciousness. Many confabulating patients do not seem to realize that there are huge gaps in their awareness. What can we learn about consciousness from this fact?

Confabulation and Sense of Self

Many cases of confabulation occur when someone is asked to provide information about him or herself. I noted in chapter 1 that one might take a Dennettian approach to confabulation by arguing that it is caused by the human need and tendency to tell a coherent "story" about ourselves, both to ourselves and others. Dennett (1991) said that we create a sense of self by creating a "center of narrative gravity," a basically fictional entity (like the center of gravity of an object), for certain social purposes. To gain the benefits of membership in a society, I must present myself to others as a unified and responsible being, capable of explaining exactly why I do what I do and say what I say. Flanagan (2000) calls the telling of such stories acts of "self-representing," but points out that the fact that this sort of self is a construction does not imply that it has no functional role.

Continuing this line of thought would involve clarifying the bewildering variety of uses of "self" employed in philosophy and psychology. Once one develops a clear notion of what the self is supposed to be, one could then examine the phenomena described in confabulating patients to determine whether they reveal the type of self-presenting, or self-projecting process at work. The one significant objection to this we have seen thus far is that this view of confabulation only explains positive confabulations about oneself. Many confabulators allege events that are negative to them, and a large percentage of confabulations are not about the patient, such as confabulatory answers given during perceptual testing.

Confabulation and Dreams

Some writers have referred to dreams as confabulations (e.g., Joseph 1996). There are interesting similarities. Dreams contain both kinds of errors present in confabulation. They initially contain ill-grounded representations, and second, the dreamer typically fails to notice the ill-groundedness of the representations. We are uncritical in dreams in a way that may be similar to the gullibility of confabulators; perhaps in both cases frontally based checking processes are not operating. We saw in chapter 3 that a cingulectomy operation can briefly produce a state in which the patient cannot discern whether she or he is dreaming or awake—vivid day-dreaming.

Are dream states conscious mental states? The question is difficult because while dreams in rapid eye-movement sleep contain vivid impressions, the conscious states involved seem dull and dark compared with waking consciousness. Perhaps when we dream one of the processes involved in producing conscious states continues to operate, while others are shut down. One conscious impression in dreams is a curious kind of familiarity. Sometimes in dreams we have a sense of the identity of a person that is contrary to that person's appearance. We say, for instance, "My parents were there, but they didn't look like my parents." This is consistent with the distinction between internal and external representations of persons (see section 5.4). In the dream cases, the internal representation is operating, and it is powerful enough to convey a sense of identity even in the face of a contrary physical appearance, just as it does in Fregoli's syndrome.

Rationality

There is also something important here for our understanding of human rationality. Are confabulators irrational? Are they behaving irrationally when they confabulate? I noted in chapter 3 that there are many examples in the literature of memory confabulators contradicting themselves—the essence of irrationality. We must think and reason in order to achieve or move nearer to our goals. This process often involves the rejection of plans or schemes; what the phenomenon of confabulation shows is that there is a natural division in the human mind between representation-creating, and representation-checking processes. Of the two errors involved in confabulation, the formation of an ill-grounded belief and the failure to know that it is ill-grounded, the second has more serious implications for a person's rationality. If a person knows that his brain tends to produce ill-grounded beliefs in a certain realm, there are many ways he can compensate for this. We do not regard such a person as irrational in any way. On the other hand, if he perpetually fails to know that he has ill-grounded beliefs, this begins to affect our opinion of his rationality.

Perhaps the most general way one could describe the function of the orbitofrontal cortex is to say that it is involved in standards—social, ethical, and otherwise. This cortex produces a feeling of revulsion at substandard products, actions, and emotions. The person with orbitofrontal damage often knows at some level that his actions are below standard, but no longer cares. His brain no longer produces the emotions powerful enough to stop him from behaving inappropriately. The idea that we inhibit contemplated actions as part of normal life has obvious ethical implications. Many of the phenomena of confabulation and disinhibition also relate strongly to our everyday notion of conscience.

Finally, if the connection between confabulation and mind reading is as intimate as I suggest in chapter 5, this communicates important clues about human nature. We internalize the standards of others by internalizing simulations of their minds, and these can work as checking and inhibiting forces in our own minds.

References

Ackil, J. K. and Zaragoza, M. S. 1998. Memorial consequences of forced confabulation: Age differences in susceptibility to false memories. *Developmental Psychology* 34: 1358–1372.

Adair, J. C., Schwartz, R. L., Duk, L. N., Fennell, E., Gilmore, R. L., and Heilman, K. M. 1997. Anosognosia: Examining the disconnection hypothesis. *Journal of Neurology, Neurosurgery, and Psychiatry* 63: 798–800.

Adolphs, R., Damasio, H., Tranel, D., and Damasio, A. R. 1996. Cortical systems for the recognition of emotion in facial expression. *Journal of Neuroscience* 16: 7678–7687.

Adolphs, R., Tranel, D., and Damasio, A. R. 2001. Neural systems subserving emotion: Lesion studies of the amygdala, somatosensory cortices, and ventromedial prefrontal cortices. In *Handbook of Neurophysiology*, ed. G. Gainotti. New York: Elsevier.

Albert, M. L. 1973. A simple test of visual neglect. *Neurology* 23: 658–664.

Alexander, M. P., Stuss, D. T., and Benson, D. F. 1979. Capgras' syndrome: A reduplicative phenomenon. *Neurology* 29: 334–339.

Amaral, D. G., Price, J. L., Pitkanen, A., and Carmichael, S. T. 1992. Anatomical organization of the primate amygdaloid complex. In *The Amygdala: Neurobiological Aspects of Emotion, Memory, and Mental Dysfunction*, ed. J. P. Aggleton. New York: Wiley–Liss.

American Psychiatric Association. 1994. *Diagnostic and Statistical Manual of Mental Disorders* (4th ed.). Washington, D.C.: American Psychiatric Press.

Anderson, S. W., Bechara, A., Damasio, H., Tranel, D., and Damasio, A. R. 1999. Impairment of social and moral behavior related to early damage in human prefrontal cortex. *Nature Neuroscience* 2: 1032–1037.

Anderson, M. W. and Green, C. 2001. Suppressing unwanted memories by executive control. *Nature* 410: 366–369.

Anton, G. 1899. Über die Selbstwahrnehmung der Herderkranken des Gehirns durch den Kranken bei Rindenblindheit und Rindentaubheit. *Archiven für Psychiatrik* 32: 86–127.

Armstrong, D. M. 1968. *A Materialist Theory of Mind*. London: Routledge and Kegan Paul.

Babinski, J. 1914. Contribution a l'etude des troubles mentaux dans l'hemiplegie organique cerebrale (anosognosie). *Revue Neurologique* (Paris) 27: 845–848.

Bachevalier, J. 1994. Medial temporal lobe structures: A review of clinical and experimental findings. *Neuropsychologia* 32: 627–648.

Bachevalier, J. 1996. Brief report: Medial temporal lobe and autism: A putative animal model in primates. *Journal of Autism and Developmental Disorders* 26: 217–220.

Bachevalier, J. and Mishkin, M. 1986. Visual recognition impairment following ventromedial but not dorsolateral prefrontal lesions in monkeys. *Behavioral Brain Research* 20: 249–261.

Bacon, A. L., Fein, D., Morris, R., Waterhouse, L., and Allen, D. 1998. The responses of autistic children to the distress of others. *Journal of Autism and Developmental Disorders* 28: 129–142.

Bagshaw, M. H., Kimble, D. P., and Pribram, K. H. 1965. The GSR of monkeys during orientation and habituation and after ablation of the amygdala, hippocampus, and inferotemporal cortex. *Neuropsychologia* 3: 111–119.

Barbizet, J. 1963. Defect of memorizing of hippocampal mamillary origin. *Journal of Neurology, Neurosurgery, and Psychiatry* 26: 127–135.

Barbas, H. 1993. Organization of cortical afferent input to the orbitofrontal area in the rhesus monkey. *Neuroscience* 56: 841–864.

Barbas, H. 2000. Proceedings of the human cerebral cortex: From gene to structure and function. *Brain Research Bulletin* 52: 319–330.

Barbas, H., Ghashghaei, H. T., Rempel-Clower, N. L., and Xiao, D. 2002. Anatomic basis of functional specialization in prefrontal cortices in primates. In *Handbook of Neuropsychology*, ed. J. Grafman. New York: Elsevier.

Barbas, H., Henion, T. H., and Dermon, C. R. 1991. Diverse thalamic projections to the prefrontal cortex in the rhesus monkey. *Journal of Comparative Neurology* 313: 65–94.

Baron-Cohen, S. 1995. *Mindblindness: An Essay on Autism and Theory of Mind.* Cambridge, Mass.: MIT Press.

Baron-Cohen, S. Baldwin, D. A., and Crowson, M. 1997. Do children with autism use the speaker's direction of gaze strategy to crack the code of language? *Child Development* 68: 48–57.

Baron-Cohen, S., Ring, H. A., Bullmore, E. T., Wheelwright, S., Ashwin, C., and Williams, S. C. 2000. The amygdala theory of autism. *Neuroscience Biobehavioral Reviews* 24: 355–364.

Baron-Cohen, S., Ring, H., Moriarty, J., Schmitz, B., Costa, D., and Ell, P. 1994. Recognition of mental state terms: Clinical findings in children with autism and a functional neuroimaging study of normal adults. *British Journal of Psychiatry* 165: 640–649.

Bauer, R. M. 1984. Autonomic recognition of names and faces in prosopagnosia: A neuropsychological application of the Guilty Knowledge Test. *Neuropsychologia* 22: 457–469.

Bauer, R. M. 1986. The cognitive psychophysiology of prosopagnosia. In *Aspects of Face Processing*, eds. H. Ellis, M. Jeeves, F. Newcombe, and A. W. Young. Dordrecht: Nijhoff.

Bauman, M. L. and Kemper, T. L. 1994. Neuroanatomic observations of the brain in autism. In *The Neurobiology of Autism*, eds. M. L. Bauman and T. L. Kemper. Baltimore: Johns Hopkins University Press.

Bear, D. M. 1983. Hemispheric specialization and the neurology of emotion. *Neurological Review* 40: 195–202.

Bechara, A., Damasio, H., and Damasio, A. R. 2000a. Emotion, decision-making and the orbitofrontal cortex. *Cerebral Cortex* 10: 295–307.

Bechara, A., Damasio, A. R., Damasio, H., and Anderson, S. W. 1994. Insensitivity to future consequences following damage to human prefrontal cortex. *Cognition* 50: 7–15.

Bechara, A., Tranel, D., Damasio, H. 2000b. Characterization of the decision-making deficit of patients with ventromedial prefrontal cortex lesions. *Brain* 123: 2189–2202.

Bechara, A., Tranel, D., Damasio, H., and Damasio, A. R. 1996. Failure to respond autonomically to anticipated future outcomes following damage to prefrontal cortex. *Cerebral Cortex* 6: 215–225.

Beeckmans, K., Vancoillie, P., and Michiels, K. 1998. Neuropsychological deficits in patients with anterior communicating artery syndrome: A multiple case study. *Acta Neurologica Belgica* 98: 266–278.

Beers, C. W. 1953. *A Mind that Found Itself: An Autobiography* (7th ed.). New York: Doubleday.

Behrmann, M. and Moscovitch, M. 2001. Face recognition: Evidence from intact and impaired performance. In *Handbook of Neuropsychology*, ed. M. Behrmann. New York: Elsevier.

Bender, L., Curran, F. J., and Schilder, P. 1916. Organization of memory traces in the Korsakoff syndrome. *Archives of Neurological Psychiatry (London)* 39: 482–487.

Benson, D. F. 1994. *The Neurology of Thinking*. Oxford: Oxford University Press.

Benson, D. F., Djenderedjian, A., Miller, B. L., Pachana, N. A., Chang, L., Itti, L., Eng, G. E., and Mena, I. 1996. Neural basis of confabulation. *Neurology* 46: 1239–1243.

Benson, D. F., Gardner, H., and Meadows, J. C. 1976. Reduplicative paramnesia. *Neurology* 26: 147–151.

Berlyne, N. 1972. Confabulation. *British Journal of Psychiatry* 120: 31–39.

Bermudez, J. L. 2000. Self-deception, intentions, and contradictory beliefs. *Analysis* 60: 309–319.

Berrios, G. E. 1998. Confabulations: A conceptual history. *Journal of the History of the Neurosciences* 7: 225–241.

Beukelman, D. R., Flowers, C. R., and Swanson, P. D. 1980. Cerebral disconnection associated with anterior communicating artery aneurysm: Implications for evaluation of symptoms. *Archives of Physical Medicine and Rehabilitation* 61: 18–22.

Bianchi, L. 1922. *The Mechanism of the Brain*. Trans. J. H. MacDonald. Edinburgh: E. and S. Livingstone.

Bichat, F. 1805. *Recherches Physiologiques sur la Vie et la Mort* (3rd ed.). Paris: Brosson/Gabon.

Bick, P. A. 1986. The syndrome of intermetamorphosis. *Bibliotheca Psychiatrica* 164: 131–135.

Bisiach, E. and Luzzatti, C. 1978. Unilateral neglect of representational space. *Cortex* 14: 129–133.

Bisiach, E. and Geminiani, G. 1991. Anosognosia related to hemiplegia and hemianopia. In *Awareness of Deficit After Brain Injury: Clinical and Theoretical Issues*, eds. G. P. Prigatano and D. L. Schacter. Oxford: Oxford University Press.

Bisiach, E., Vallar, G., Perani, D., Papagno, C., and Berti, A. 1986. Unawareness of disease following lesions of the right hemisphere: Anosognosia for hemiplegia and anosognosia for hemianopia. *Neuropsychologia* 24: 471–482.

Blair, R. J. R., Colledge, E., and Mitchell, D. G. V. 2001. Somatic markers and response reversal: Is there orbitofrontal cortex dysfunction in boys with psychopathic tendencies? *Journal of Abnormal Child Psychology* 29: 499–511.

Blanke, O., Ortigue, S., Landis, T., and Seeck, M. 2002. Stimulating illusory own-body perceptions. *Nature* 419: 269–270.

Bleuler, E. P. 1924. *Textbook of Psychiatry*. New York: Macmillan.

Blumer, D. and Benson, D. F. 1975. Personality changes with frontal and temporal lobe lesions. In *Psychiatric Aspects of Neurologic Disease*, eds. D. F. Benson and D. Blumer. New York: Grune and Stratton.

Bogen, J. E. and Vogel, P. J. 1962. Cerebral commissurotomy in man. *Bulletin of the Los Angeles Neurological Society* 27: 169–172.

Bonhoeffer, K. 1901. *Die Akuten Geisteskrankehiten der Gewohnheitstrinker*, Jena: Gustav Fischer.

Bonhoeffer, K. 1904. Der Korsakowische Symptomcomplex in seinen Beziehungen zu den verschieden Krankheitsformen. *Allgemeine Zeitschrift für Psychiatrie* 61: 744–752.

Box, O., Laing, H., and Kopelman, M. 1999. The evolution of spontaneous confabulation, delusional misidentification and a related delusion in a case of severe head injury. *Neurocase* 5: 251–262.

Breen, N., Caine, D., and Coltheart, M. 2000a. Models of face recognition and delusional misidentification: A critical review. *Cognitive Neuropsychology* 17: 55–71.

Breen, N., Caine, D., Coltheart, M., Hendy, J., and Roberts, C. 2000b. Towards an understanding of delusions of misidentification: Four case studies. In *Pathologies of Belief*, eds. M. Coltheart and M. Davies. Oxford: Blackwell.

Breiter, H. C., Etcoff, N. L., Whalen, P. J., Kennedy, W. A., and Rauch, S. L. et al. 1996. Response and habituation of the human amygdala during visual processing of facial expressions. *Neuron* 17: 875–887.

Brion, S., Pragier, C., Guerin, R., and Teitgen, M. M. C. 1969. Korsakoff syndrome due to bilateral softening of the fornix. *Revue Neurologique* 120: 255–262.

Brothers, L. 1989. A biological perspective on empathy. *American Journal of Psychiatry* 146: 10–19.

Brothers, L. 1990. The social brain: A project for integrating primate behavior and neurophysiology in a new domain. *Concepts in Neuroscience* 1: 27–51.

Brothers, L. 1995. Neurophysiology of the perception of intentions by primates. In *The Cognitive Neurosciences*, ed. M. S. Gazzaniga. Cambridge, Mass.: MIT Press.

Brothers, L. and Ring, B. 1992. A neuroethological framework for the representation of minds. *Journal of Cognitive Neuroscience* 4: 107–118.

Budson, A. E., Roth, H. L., Rentz, D. M., and Ronthal, M. 2000. Disruption of the ventral visual stream in a case of reduplicative paramnesia. *Annals of the New York Academy of Sciences* 911: 447–452.

Burgess, P. W. and Shallice, T. 1996. Confabulation and the control of recollection. *Memory* 4: 359–411.

Butler, P. V. 2000. Diurnal variation in Cotard's syndrome (copresent with Capgras' delusion) following traumatic brain injury. *Australian and New Zealand Journal of Psychiatry* 34: 684–687.

Cabeza, R., Rao, S. M., Wagner, A. D., Mayer, A. R., and Schacter, D. L. 2001. Can medial temporal lobe regions distinguish true from false? An event-related functional MRI study of veridical and illusory recognition memory. *Proceedings of the National Academy of Sciences U.S.A.* 98: 4805–4810.

Cameron, N. and Margaret, A. 1951. *Behavior Pathology*. Cambridge, Mass.: Houghton Mifflin.

Campbell, R., Heywood, C. A., Cowey, A., Regard, M., and Landis, T. 1990. Sensitivity to eye gaze in prosopagnosic patients and monkeys with superior temporal sulcus ablation. *Neuropsychologia* 28: 1123–1142.

Capgras, J. and Reboul-Lachaux, J. 1923. L'illusion des "sosies" dans un délire systématisé chronique. *Bulletin de la Société Clinique de Médecine Mentale* 11: 6–16.

Cappa, S. F., Sterzi, R., Vallar, G., and Bisiach, E. 1987. Remission of hemineglect and anosognosia after vestibular stimulation. *Neuropsychologia* 25: 775–782.

Carey, D. P., Perrett, D. I., and Oram, M. W. 1997. Recognizing, understanding, and reproducing action. In *Handbook of Neuropsychology*, vol. XI, eds. F. Boller and J. Grafman. New York: Elsevier.

Carmichael, S. T. and Price, J. L. 1995a. Limbic connections of the orbital and medial prefrontal cortex in macaque monkeys. *Journal of Comparative Neurology* 363: 615–641.

Carmichael, S. T. and Price, J. L. 1995b. Sensory and premotor connections of the orbital and medial prefrontal cortex of macaque monkeys. *Journal of Comparative Neurology* 363: 642–664.

Carmichael, S. T. and Price, J. L. 1996. Connectional networks within the orbital and medial prefrontal cortex of macaque monkeys. *Journal of Comparative Neurology* 371: 179–207.

Carpenter, M. B. 1985. *Core Text of Neuroanatomy*. Baltimore: Williams and Wilkins.

Carter, C. S., Braver, T. S., Barch, D. M., Botvinick, M. M., Noll, D., and Cohen, J. D. 1998. Anterior cingulate cortex, error detection, and the online monitoring of performance. *Science* 280: 747–749.

Case, R. 1992. The role of the frontal lobes in the regulation of cognitive development. *Brain and Cognition* 20: 51–73.

Cavada, C., Company, T., Tejedor, J., Cruz-Rizzolo, R. J., and Reinoso-Suarez, F. 2000. The anatomical connections of the macaque monkey orbitrofrontal cortex. A review. *Cerebral Cortex* 10: 220–242.

Ceci, S. J., Huffman, M. L., and Smith E. 1994. Repeatedly thinking about a non-event: Source misattributions among preschoolers. *Consciousness and Cognition* 3: 388–407.

Chao, L. L., Martin, A., and Haxby, J. V. 1999. Are face-responsive regions selective only for faces? *Neuroreport* 10: 2945–2950.

Chatterjee, A. 1995. Cross-over, completion and confabulation in unilateral spatial neglect. *Brain* 118: 455–465.

Christodoulou, G. N. 1977. The syndrome of Capgras. *British Journal of Psychiatry* 130: 556–564.

Churchland, P. M. 1979. *Scientific Realism and the Plasticity of Mind*. Cambridge: Cambridge University Press.

Churchland, P. M. 1986. The continuity of philosophy and the sciences. *Mind and Language* 1: 5–14.

Churchland, P. S. 1986. *Neurophilosophy: Toward a Unified Science of the Mind/Brain*. Cambridge, Mass.: MIT Press.

Churchland, P. S. and Sejnowski, T. 1992. *The Computational Brain*. Cambridge, Mass.: MIT Press.

Cohen, N. J. and Squire, L. R. 1980. Preserved learning and retention of pattern-analyzing skill in amnesia: Dissociation of knowing how and knowing that. *Science* 210: 207–210.

Cook, N. D. 1984. Homotopic callosal inhibition. *Brain and Language* 23: 116–125.

Courbon, P. and Fail, G. 1927. Syndrome d'illusion de Fregoli et schizophrenie. *Bulletin de la Société Clinique de Médecine Mentale* 20: 121–125.

Courbon, P. and Tusques, J. 1932. L'illusion d'intermetamorphose et de charme. *Annals of Medical Psychology* 90: 401–406.

Critchley, M. 1953. *The Parietal Lobes*. London: Edward Arnold.

Critchley, H. D., Elliott, R., Mathias, C. J., and Dolan, R. J. 2000. Neural activity relating to generation and representation of galvanic skin conductance responses: A functional magnetic resonance imaging study. *Journal of Neuroscience* 20: 3033–3040.

Critchley, H. D., Mathias, C. J., and Dolan, R. J. 2001. Neural activity in the human brain relating to uncertainty and arousal during anticipation. *Neuron* 29: 537–545.

Cummings, J. L. 1985. *Clinical Neuropsychiatry*. Orlando, Fla.: Grune and Stratton.

Cunningham, J. M., Pliskin, N. H., Cassisi, J. E., Tsang, B., and Rao, S. M. 1997. Relationship between confabulation and measures of memory and executive function. *Journal of Clinical and Experimental Neuropsychology* 19: 867–877.

Curran, T., Schachter, D. L., Norman, K. A., and Galluccio, L. 1997. False recognition after a right frontal lobe infarction: Memory for general and specific information. *Neuropsychologia* 35: 1035–1049.

Cutting, J. 1978. Relationship between Korsakoff's syndrome and "alcoholic dementia." *British Journal of Psychiatry* 132: 240–251.

Cutting, J. 1991. Delusional misidentification and the role of the right hemisphere in the appreciation of identity. *British Journal of Psychiatry* 159: 70–75.

Dalla Barba, G. 1993. Different patterns of confabulation. *Cortex* 29: 567–581.

Dalla Barba, G., Cipoletti, L., and Denes, G. 1990. Autobiographical memory loss and confabulation in Korsakoff's syndrome: A case report. *Cortex* 26: 525–534.

Damasio, A. R. 1994. *Descartes' Error*. New York: G. P. Putnam.

Damasio, A. R. 1996. The somatic marker hypothesis and the possible functions of the prefrontal cortex. *Philosophical Transactions of the Royal Society of London, Series B* 351: 1413–1420.

Damasio, A. R. 1999. *The Feeling of What Happens: Body and Emotion in the Making of Consciousness*. New York: Harcourt Brace.

Damasio, A. R. 2000. A neural basis for sociopathy. *Archives of General Psychiatry* 57: 128–129.

Davidson, D. 1986. Deception and division. In *The Multiple Self*, ed. J. Elster. Cambridge: Cambridge University Press.

Davidson, D. 1998. Who is fooled? In *Self-Deception and Paradoxes of Rationality*, ed. J.-P. Dupuy. Stanford, Calif.: CSLI Publications.

Davidson, R. A., Fedio, P., Smith, B. D., Aureille, E., and Martin, A. 1992. Lateralized mediation of arousal and habituation: Differential bilateral electrodermal activity in unilateral temporal lobectomy patients. *Neuropsychologia* 30: 1053–1063.

Deese, J. 1959. On the prediction of occurrence of particular verbal intrusions in immediate recall. *Journal of Experimental Psychology* 58: 17–22.

Deglin, V. L. Kinsbourne, M. 1996. Divergent thinking styles of the hemispheres: How syllogisms are solved during transitory hemispheric suppression. *Brain and Cognition* 31: 285–307.

DeHaan, E. H. F., Bauer, R., and Greve, K. W. 1992. Behavioral and physiological evidence for covert recognition in a prosopagnosic patient. *Cortex* 28: 77–95.

Delbecq-Derousne, J., Beauvois, M. F., and Shallice, T. 1990. Preserved recall versus impaired recognition. *Brain* 113: 1045–1074.

Delgado, J. M. R. 1969. *Physical Control of the Mind: Toward a Psychocivilized Society*. New York: Harper and Row.

DeLuca, J. 1993. Predicting neurobehavioral patterns following anterior communicating artery lesion. *Cortex* 29: 639–647.

DeLuca, J. 2001. A cognitive neuroscience perspective on confabulation. *Neuro-Psychoanalysis* 2: 119–132.

DeLuca, J. and Cicerone, K. D. 1991. Confabulation following aneurysm of the anterior communicating artery. *Cortex* 27: 417–423.

DeLuca, J. and Diamond, B. J. 1995. Aneurysm of the anterior communicating artery: A review of the neuroanatomical and neuropsychological sequelae. *Journal of Clinical and Experimental Neuropsychology* 17: 100–121.

Demeter, S., Rosene, D. L., and Van Hoesen, G. W. 1990. Fields of origin and pathways of the interhemispheric commissures in the temporal lobes of macaques. *Journal of Comparative Neuroscience* 302: 29–53.

Dengerink, H. A. and Bertilson, H. S. 1975. Psychopathy and physiological arousal in an aggressive task. *Psychophysiology* 12: 682–684.

Dennett, D. 1991. *Consciousness Explained*. Boston: Little, Brown.

De Pauw, K. W. and Szulecka, T. K. 1988. Dangerous delusions: Violence and the misidentification syndromes. *British Journal of Psychiatry* 152: 91–96.

De Pauw, K. W., Szulecka, T. K., and Poltock, T. L. 1987. Fregoli syndrome after cerebral infarction. *Journal of Nervous and Mental Disease* 175: 433–438.

Descartes, R. 1641/1967. *Meditations on First Philosophy*. Trans. E. Haldane and D. Ross. Cambridge: Cambridge University Press.

Devinsky, O., Morrell, M. J., and Vogt, B. A. 1995. Contributions of the anterior cingulate to behavior. *Brain* 118: 279–306.

DiVirgilio, G., Clarke, S., Pizzolato, G., and Schaffner, T. 1999. Cortical regions contributing to the anterior commissure in man. *Experimental Brain Research* 124: 1–7.

Drake, M. E. 1987. Postictal Capgras' syndrome. *Clinical Neurology and Neurosurgery* 89: 271–274.

Dywan, J. 1995. The illusion of familiarity: An alternative to the report-criterion account of hypnotic recall. *International Journal of Clinical and Experimental Hypnosis* 43: 194–211.

Dywan, J. 1998. Toward a neurophysiological model of hypnotic memory effects. *American Journal of Clinical Hypnosis* 40: 217–230.

Edelstyn, N. M. and Oyebode, F. 1999. A review of the phenomenology and cognitive neuropsychological origins of the Capgras syndrome. *International Journal of Geriatric Psychiatry* 14: 48–59.

Edelstyn, N. M., Oyebode, F., and Barrett, K. 2001. The delusions of Capgras and intermetamorphosis in a patient with right-hemisphere white-matter pathology. *Psychopathology* 34: 299–304.

Ehrt, U. 1999. A zoocentric Capgras' syndrome. *Psychiatric Praxis* 26: 43–44.

Ekman, P. 1992. *Telling Lies: Clues to Deceit in the Marketplace, Politics, and Marriage*. New York: W. W. Norton.

Elliot, R., Rees. G., and Dolan, R. J. 1999. Ventromedial prefrontal cortex mediates guessing. *Neuropsychologia* 37: 403–411.

Ellis, H. D. 1994. The role of the right hemisphere in the Capgras' delusion. *Psychology* 27: 177–185.

Ellis, H. D., Luauté, J. P., and Retterstol, N. 1994. Delusional misidentification syndromes. *Psychopathology* 27: 117–120.

Ellis, H. D. and Young, A. W. 1990. Accounting for delusional misidentifications. *British Journal of Psychiatry* 157: 239–248.

Ellis, H. D., Young, A. W., Quayle, A. H., and de Pauw, K. W. 1997. Reduced autonomic responses to face in Capgras' delusion. *Proceedings of the Royal Society of London, Series B* 264: 1085–1092.

Farah, M. J. 1994. Visual perception and visual awareness after brain damage: A tutorial overview. In *Attention and Performance XV: Conscious and Nonconscious Information Processing*, eds. M. Moscovitch and C. Umilta. Cambridge, Mass.: MIT Press.

Feinberg, T. E. 2001. *Altered Egos: How the Brain Creates the Self*. Oxford: Oxford University Press.

Feinberg, T. E., Haber, L. D., and Leeds, N. E. 1990. Verbal asomatognosia. *Neurology* 40: 1391–1394.

Feinberg, T. E. and Roane, D. M. 1997. Anosognosia, completion and confabulation: The neutral-personal dichotomy. *Neurocase* 3: 73–85.

Feinberg, T. E., Roane, D. M., Kwan, P. C., Schindler, R. J., and Haber, L. D. 1994. Anosognosia and visuoverbal confabulation. *Archives of Neurology* 31: 468–473.

Fingarette, H. 1998. Self-deception needs no explaining. *Philosophical Quarterly* 48: 289–301.

Finger, S. 1994. *Origins of Neuroscience: A History of Explorations into Brain Function*. Oxford: Oxford University Press.

Fischer, R. S., Alexander, M. P., D'Esposito, M. D., and Otto, R. 1995. Neuropsychological and neuroanatomical correlates of confabulation. *Journal of Clinical and Experimental Neuropsychology* 17: 20–28.

Flanagan, O. 1996. *Self-Expressions: Mind, Morals, and the Meaning of Life.* Oxford: Oxford University Press.

Flanagan, O. 2000. *Dreaming Souls: Sleep, Dreams, and the Evolution of the Conscious Mind.* Oxford: Oxford University Press.

Fleminger, S. and Burns, A. 1993. The delusional misidentification syndromes in patients with and without evidence of organic cerebral disorder: A structured review of case reports. *Biological Psychiatry* 33: 22–32.

Förstl, H., Almeida, O. P., Owen, A., Burns, A., and Howard, R. 1991. Psychiatric, neurological and medical aspects of misidentification syndromes: A review of 260 cases. *Psychological Medicine* 21: 905–950.

Förstl, H. and Beats, B. 1992. Charles Bonnet's description of Cotard's delusion and reduplicative paramnesia in an elderly patient (1788). *British Journal of Psychiatry* 160: 416–418.

Fowles, D. C. 1980. The three arousal model: Implications of Gray's two-factor theory for heart rate, electrodermal activity, and psychopathy. *Psychophysiology* 17: 87–104.

Fowles, D. C. 2000. Electrodermal hyporeactivity and antisocial behavior: Does anxiety mediate the relationship? *Journal of Affective Disorders* 61: 177–179.

Frazer, S. J. and Roberts, J. M. 1994. Three cases of Capgras' syndrome. *British Journal of Psychiatry* 164: 557–559.

Freeman, W. and Watts, J. W. 1942. *Psychosurgery, Intelligence, Emotions, and Social Behavior Following Prefrontal Lobotomy for Mental Disorders.* Springfield, Ill.: Charles C. Thomas.

Freedman, A. M., Kaplan, H. I., and Sadock, B. J. 1975. *Comprehensive Textbook of Psychiatry,* Baltimore: Williams and Wilkins.

Freud, S. 1938. *The Interpretation of Dreams.* In *The Basic Writings of Sigmund Freud.* Trans. A. A. Brill. New York: Random House. Originally published in 1900.

Fried, I., Mateer, C., Ojemann, G., Wohns, R., and Fedio, P. 1982. Organization of visuospatial functions in human cortex: Evidence from electrical stimulation. *Brain* 105: 349–371.

Frith, C. D. 1992. *The Cognitive Neuropsychology of Schizophrenia.* Hove, U.K.: Lawrence Erlbaum.

Fukutake, T., Akaba, K., Ito, S., Okuba, T. 2002. Severe personality changes after unilateral left paramedian thalamic infarct. *European Neurology* 47: 156–160.

Fulford, K. W. M. 1989. *Moral Theory and Medical Practice*. Cambridge: Cambridge University Press.

Fulton, J. F. and Bailey, P. 1929. Tumors in the region of the third ventricle: Their diagnosis and relation to pathological sleep. *Journal of Nervous and Mental Disorders* 69:1–25, 145–164.

Funnell, M. G., Corballis, P. M., and Gazzaniga, M. S. 2000. Hemispheric interactions and specialization: Insight from the split brain. In *Handbook of Neuropsychology* (2nd ed.). vol. 1, eds. F. Boller, J. Grafman, and G. Rizzolatti. New York: Elsevier.

Fuster, J. M. 1995. *Memory in the Cerebral Cortex*. Cambridge, Mass.: MIT Press.

Gaffan, D. and Murray, E. 1990. Amygdalar interaction with the mediodorsal nucleus of the thalamus and the ventromedial prefrontal cortex in stimulus-reward associative learning in the monkey. *Journal of Neuroscience* 10: 3479–3493.

Gaffan, D., Parker, A., Easton, A. 2001. Dense amnesia in the monkey after transection of fornix, amygdala and anterior temporal stem. *Neuropsychologia* 39: 51–70.

Gainotti, G. 1972. Emotional behavior and hemispheric side of lesion. *Cortex* 8: 41–55.

Gainotti, G. 1975. Confabulation of denial in senile dementia. *Psychiatria Clinica* 8: 99–108.

Gainotti, G. 2001. Components and levels of emotion disrupted in patients with unilateral brain damage. In *Handbook of Neuropsychology* (2nd ed.). vol. 5, ed. G. Gainotti. New York: Elsevier.

Gamper, E. 1928. Schlaff—Delirium tremens—Korsakawsches. *Zentralblatt für Neurologie* 51: 236–239.

Gardner, H. 1974. *The Shattered Mind*. New York: Vintage Books.

Gassell, M. M. and Williams, D. 1963. Visual function in patients with homonymous hemianopia. III. The completion phenomenon; insight and attitude to the defect; and visual functional efficiency. *Brain* 86: 229–260.

Gazzaniga, M. S. 1983. Right hemisphere language following brain bisection: A Twenty Year Perspective. *American Psychologist* 38: 525–537.

Gazzaniga, M. S. 1992. *Nature's Mind*. New York: Basic Books.

Gazzaniga, M. S. 1995a. Principles of human brain organization derived from split-brain studies. *Neuron* 14: 217–228.

Gazzaniga, M. S. 1995b. *Consciousness and the Cerebral Hemispheres*. In *The Cognitive Neurosciences*, ed. M. S. Gazzaniga. Cambridge, Mass.: MIT Press.

Gazzaniga, M. S. 1998. *The Mind's Past*. Berkeley, Calif.: University of California Press.

Gazzaniga, M. S. 2000. Cerebral specialization and interhemispheric communication: Does the corpus callosum enable the human condition? *Brain* 123: 1293–1326.

Gazzaniga, M. S., LeDoux, J. E., and Wilson, D. H. 1977. Language, praxis, and the right hemisphere: Clues to some mechanisms of consciousness. *Neurology* 27: 1144–1147.

Gazzaniga, M. S. and LeDoux, J. E. 1978. *The Integrated Mind*. New York: Plenum Press.

Gazzaniga, M. S., Ivry, R., and Mangun, G. R. 2002. *Cognitive Neuroscience: The Biology of the Mind* (2nd ed.). London: W. W. Norton and Company, Ltd.

George, M. S., Costa, D. C., Kouris, K., Ring, H., and Ell, P. 1992. Cerebral blood flow abnormalities in adults with infantile autism. *Journal of Nervous and Mental Disease* 180: 413–417.

Geschwind, N. 1965. Disconnexion syndromes in animals and man. *Brain* 88: 237–644.

Gianetti, E. 1997. *Lies We Live By: The Art of Self-Deception*. London: Bloomsbury.

Gilovich, T. 1991. *How We Know What Isn't So*. New York: Macmillan.

Gloor, P. 1986. The role of the human limbic system in perception, memory and affect: Lessons from temporal lobe epilepsy. *In The Limbic System: Functional Organization and Clinical Disorders*, eds. B. K. Doane and K. E. Livingston. New York: Raven.

Gloor, P. 1997. *The Temporal Lobe and the Limbic System*. Oxford: Oxford University Press.

Gloor, P., Salanova, V., Olivier, A., and Quesney, L. F. 1993. The human dorsal hippocampal commissure. An anatomically identifiable and functional pathway. *Brain* 116: 1249–1273.

Goddard, G. V., McIntyre, D. C., and Leech, C. K. 1969. A permanent change in brain function resulting from daily electrical stimulation. *Experimental Neurology* 25: 295–330.

Goktepe, E. O., Young, L. B., and Bridges, P. K. 1975. A further review of the results of stereotactic subcaudate tractotomy. *British Journal of Psychiatry* 126: 270–280.

Goldman, A. 1986. *Epistemology and Cognition*. Cambridge, Mass.: Harvard University Press.

Goldman, A. 1998. Justification externalism: Pro and con. *Routledge Encyclopedia of Philosophy*. London: Routledge.

Goldman-Rakic, P. S. 1987. Circuitry of primate prefrontal cortex and regulation of behavior by representational memory. In *Handbook of Physiology, The Nervous System, Higher Functions of the Brain*, ed. F. Plum. Bethesda, Md.: American Physiological Society.

Grattan, L. M. and Eslinger, P. J. 1989. Higher cognition and social behavior: Changes in cognitive flexibility and empathy after cerebral lesions. *Neuropsychology* 3: 175–185.

Grattan, L. M., Bloomer, R. H., Archambault, F. X., and Eslinger, P. J. 1994. Cognitive flexibility and empathy after cerebral lesions. *Neuropsychiatry, Neuropsychology, and Behavioral Neuroscience* 7: 251–259.

Greenblatt, M., Arnot, R., and Solomon, H. C. 1950. *Studies in Lobotomy*. New York: Grune and Stratton.

Gregory, C., Lough, S., Stone, V., Erzinclioglu, S., Martin, L., Baron-Cohen, S., and Hodges, J. R. 2002. Theory of mind in patients with frontal variant frontotemporal dementia and Alzheimer's disease: Theoretical and practical implications. *Brain* 125: 752–764.

Hakim, H., Verma, N. P., and Greiffenstein, M. F. 1988. Pathogenesis of reduplicative paramnesia. *Journal of Neurology, Neurosurgery, and Psychiatry* 51: 839–841.

Halgren, E. 1992. Emotional neurophysiology of the amygdala within the context of human cognition. In *The Amygdala: Neurobiological Aspects of Emotion, Memory, and Mental Dysfunction*, ed. J. P. Aggleton. New York: Wiley.

Halgren, E. and Marinkovic, K. 1995. Neurophysicological networks integrating human emotions. In *The Cognitive Neurosciences*, ed. M. S. Gazzaniga. Cambridge, Mass.: MIT Press.

Harding, A., Halliday, G., Caine, D., and Kril, J. 2000. Degeneration of anterior thalamic nuclei differentiates alcoholics with amnesia. *Brain* 123: 141–154.

Hare, R. D. 1972. Psychopathy and physiological response to adrenalin. *Journal of Abnormal Psychology* 79: 138–147.

Hare, R. D. 1993. *Without Conscience: The Disturbing World of the Psychopaths Among Us*. New York: Guilford Press.

Hare, R. D. and Quinn, M. J. 1971. Psychopathy and autonomic conditioning. *Journal of Abnormal Psychology* 77: 223–235.

Harlow, J. M. 1869. Recovery from the passage of an iron bar through the head. Boston: David Clapp and Sons Medical and Surgical Office. Reprinted in *History of Psychiatry* 4(1993): 271–281.

Hartmann, J. A., Wolz, W. A., Roeltgen, D. P., and Loverso, F. L. 1991. Denial of visual perception. *Brain and Cognition* 16: 29–40.

Harwood, D. G., Barker, W. W., Owndy, R. L., and Duara, R. 1999. Prevalence and correlates of Capgras' syndrome in Alzheimer's disease. *International Journal of Geriatric Psychiatry* 14: 415–420.

Hasselmo, M., Rolls, E. T., and Baylis, G. C. 1989. The role of expression and identity in the face-selective responses of neurons in the temporal visual cortex of the monkey. *Behavioral Brain Research* 32: 203–218.

Haxby, J. V., Hoffman, E. A., and Gobbini, M. I. 2000. The distributed human neural system for face perception. *Trends in Cognitive Sciences* 4: 223–233.

Hecaen, H., Talairach, J., David, M., and Dell, M. B. 1949. Coagulations limitees du thalamus dans les algies du syndrome thalamique: Resultats therapeutiques et physiologiques. *Revue Neurologique (Paris)* 81: 917–931.

Hecaen, H. and Albert, M. L. 1975. Disorders of mental functioning related to frontal lobe pathology. In *Psychiatric Aspects of Neurologic Disease*, eds. D. F. Benson and D. Blumer. New York: Grune and Stratton.

Hecaen, H. and Albert, M. 1978. *Human Neuropsychology*. New York: Wiley.

Heilman, K. M. 1991. Anosognosia: Possible neuropsychological mechanisms. In *Awareness of Deficit after Brain Injury: Clinical and Theoretical Issues*, eds. G. P. Prigatano and D. L. Schacter. Oxford: Oxford University Press.

Heilman, K. M., Barrett, A. M., and Adair, J. C. 1998. Possible mechanisms of anosognosia: A defect in self-awareness. *Philosophical Transactions of the Royal Society of London, Series B* 353: 1903–1909.

Heilman, K. M., Schwartz, H. D., and Watson, R. 1978. Hypoarousal in patients with the neglect syndrome and emotional indifference. *Neurology* 28: 229–232.

Heilman, K. M., Watson, R. T., and Valenstein, E. 1991. Neglect and related disorders. In *Clinical Neuropsychology*, eds. K. M. Heilman and E. Valentstein. Oxford: Oxford University Press.

Hellige, J. B. 1993. *Hemispheric Asymmetry: What's Right and What's Left*. Cambridge, Mass.: Harvard University Press.

Hirstein, W. 2000. Self-deception and confabulation. *Philosophy of Science, Proceedings* 67: 418–429.

Hirstein, W. 2004. *On the Churchlands*. Belmont, Calif.: Wadsworth.

Hirstein, W., Iversen, P., and Ramachandran, V. S. 2001. Autonomic responses of autistic children to people and objects. *Proceedings of the Royal Society of London, Series B* 268: 1883–1888.

Hirstein, W. and Ramachandran, V. S. 1997. Capgras' syndrome: A novel probe for understanding the neural representation of the identity and familiarity of persons. *Proceedings of the Royal Society of London, Series B* 264: 437–444.

Hoffman, E. A. and Haxby, J. V. 2000. Distinct representations of eye gaze and identity in the distributed human neural system for face perception. *Nature Neuroscience* 3: 80–84.

Holbourn, A. H. S. 1943. Mechanics of head injury. *Lancet* 2: 438–441.

Hopkins, M. J., Dywan, J., and Segalowitz, S. J. 2002. Altered electrodermal response to facial expression after closed head injury. *Brain Injury* 16: 245–257.

Hornak, J., Rolls, E. T., and Wade, D. 1996. Face and voice expression identification in patients with emotional and behavioural changes following ventral frontal lobe damage. *Neuropsychologia* 34: 247–261.

Hudson, A. J. and Grace, G. M. 2000. Misidentification syndromes related to face-specific area in the fusiform gyrus. *Journal of Neurology, Neurosurgery and Psychiatry* 69: 645–648.

Jackson, J. H. 1881. Remarks on the dissolution of the nervous system as exemplified by certain post-epileptic conditions. *Medical Press and Circular* 2: 401–432.

Jackson, J. H. 1887. Remarks on the evolution and dissolution of the nervous system. In *Selected Writings of John Hughlings Jackson*, vol. 2, ed. J. Taylor, London: Staples Press, 1932.

James, 1890. *The Principles of Psychology*. New York: Henry Holt.

Jernigan, T. L., Butters, N., DiTraglia, G., Schafer, K., Smith, T., Irwin, M., Grant, I., Schukeit, M., and Cermak, L. S. 1991. Reduced cerebral gray matter observed in alcoholics using magnetic resonance imaging. *Alcoholism: Clinical and Experimental Research* 15: 418–427.

Jeste, D. V., Galasko, D., Corey-Bloom, J., Walens, S., and Granholm, E. 1996. Neuropsychiatric aspects of the schizophrenias. In *Neuropsychiatry*, eds. B. S. Fogel, R. B. Schiffer, and S. M. Rao. Baltimore: Williams and Wilkins.

Jha, A. P., Kroll, N. E. A., Baynes, K., and Gazzaniga, M. 1997. Memory encoding following complete callosotomy. *Journal of Cognitive Neuroscience* 9: 143–159.

Johnson, C. N. and Wellman, H. M. 1980. Children's developing understanding of mental verbs: Remember, know and guess. *Child Development* 51: 1095–1102.

Johnson, M. K. 1991. Reality monitoring: Evidence from confabulation in organic brain disease patients. In *Awareness of Deficit after Brain Injury: Clinical and Theoretical Issues*, eds. G. P. Prigatano and D. L. Schacter. Oxford: Oxford University Press.

Johnson, M. K., Hayes, S. M., D'Esposito, M. D., and Raye, C. L. 2000. Confabulation. In *Handbook of Neuropsychology*, eds. J. Grafman and F. Boller. New York: Elsevier.

Johnson, M. K., Kounios, J., and Nolde, S. F. 1996. Electrophysiological brain activity and memory source monitoring. *Neuroreport* 7: 2929–2932.

Johnson, M. K., O'Connor, M., and Cantor, J. 1997. Confabulation, memory deficits, and frontal dysfunction. *Brain and Cognition* 34: 189–206.

Johnson, M. K. and Raye, C. L. 1998. False memories and confabulation. *Trends in Cognitive Sciences* 2: 137–145.

Jolly, F. 1897. Über die psychischen Störungen bei Polyneuritis. *Charite Annalen* 22: 579–612.

Jones, B. and Mishkin, M. 1972. Limbic lesions and the problem of stimulus-reinforcement. *Experimental Neurology* 36: 362–377.

Jones, H. E. 1950. The study of patterns of emotional expression. In *Feelings and Emotions: The Mooseheart Symposium*, ed. M. L. Reymert. New York: McGraw Hill.

Joseph, A. B., O'Leary, D. H., and Wheeler, H. G. 1990. Bilateral atrophy of the frontal and temporal lobes in schizophrenic patients with Capgras' syndrome: A case-control study using computed tomography. *Journal of Clinical Psychiatry* 51: 322–325.

Joseph, R. 1986. Confabulation and delusional denial: Frontal lobe and lateralized influences. *Journal of Clinical Psychology* 42: 507–520.

Joseph, R. 1996. *Neuropsychiatry, Neuropsychology, and Clinical Neuroscience: Emotion, Evolution, Cognition, Language, Memory, Brain Damage, and Abnormal Behavior*. Baltimore: Williams and Wilkins.

Joseph, R. M. 1998. Intention and knowledge in preschoolers' conception of pretend. *Child Development* 69: 966–980.

Joslyn, D., Grundivg, J. K., and Chamberlain, C. J. 1978. Predicting confabulation from the Graham-Kendall Memory-for-Design Test. *Journal of Consulting and Clinical Psychology* 46: 181–182.

Kaada, B. R., Pribram, K. H., and Epstein, J. A. 1949. Respiratory and vascular responses in monkeys from temporal pole, insula, orbital surface, and cingulate gyrus. *Journal of Neurophysiology* 12: 347–356.

Kapur, N. and Coughlan, A. K. 1980. Confabulation and frontal lobe dysfunction. *Journal of Neurology, Neurosurgery and Psychiatry* 43: 461–463.

Kapur, N., Turner, A., and King, C. 1988. Reduplicative paramnesia: Possible anatomical and neuropsychological mechanisms. *Journal of Neurology, Neurosurgery and Psychiatry* 51: 579–581.

Kartsounis, L. D., Poynton, A., Bridges, P. K., and Bartlett, J. R. 1991. Neuropsychological correlates of stereotactic subcaudate tractotomy. *Brain* 114: 2657–2673.

Kern, R. S., Van Gorp, W. G., Cummings, J., Brown, W., and Osato, S. 1992. Confabulation in Alzheimer's disease. *Brain and Cognition* 19: 172–182.

Kinsbourne, M. 1975. The mechanism of hemispheric control of the lateral gradient of attention. In *Attention and Performance*, vol. 5, eds. P. M. A. Rabbitt and S. Dornic. New York: Academic Press.

Kinsbourne, M. 1988. Integrated field theory of consciousness. In *Consciousness in Contemporary Science*, eds. A. J. Marcel and E. Bisiach. Oxford: Clarendon Press.

Kinsbourne, M. 1993. Integrated cortical field model of consciousness. In *Experimental and Theoretical Studies of Consciousness* (Ciba Foundation symposium). New York: Wiley.

Kinsbourne, M. 1995. Awareness of one's own body: An attentional theory of its nature development, and brain basis. In *The Body and the Self*, eds. J. L. Bermudez, A. Marcel, and N. Eilan. Cambridge, Mass.: MIT Press.

Kolakowski, L. 1985. *Bergson*. Oxford: Oxford University Press.

Kopelman, M. D. 1987. Two types of confabulation. *Journal of Neurology, Neurosurgery, and Psychiatry* 50: 1482–1487.

Kopelman, M. D. 1989. Remote and autobiographical memory, temporal context memory and frontal atrophy in Korsakoff and Alzheimer's patients. *Neuropsychologia* 27: 437–460.

Kopelman, M. D. 1991. Frontal dysfunction and memory deficits in the alcoholic Korsakoff syndrome and Alzheimer-type dementia. *Brain* 116: 117–137.

Kopelman, M. D. and Stanhope, N. 1997. Rates of forgetting in organic amnesia following temporal lobe, diencephalic, or frontal lobe lesions. *Neuropsychology* 11: 343–356.

Kopelman, M. D., Guinan, E. M., and Lewis, P. D. R. 1995. Delusional memory, confabulation, and frontal lobe dysfunction: A case study in DeClerambault's syndrome. *Neurocase* 1: 71–77.

Kopelman, M. D., Ng, N., and Van Den Brouke, O. 1997. Confabulation extending across episodic, personal, and general semantic memory. *Cognitive Neuropsychology* 14: 683–712.

Korsakoff, S. S. 1889. Psychic disturbance in conjunction with peripheral neuritis. Trans. M. Victor and P. I. Yakovlev. *Neurology* 5(1955): 394–406.

Korsakoff, S. S. 1892. Erinnerungsteuschungen (pseudoreminiscenzen) bei polyneuritischer psychose. *Allgemeine Zeitschrift für Psychiatrie* 47: 390–410.

Kosslyn, S. M. 1987. Seeing and imagining in the cerebral hemispheres. *Psychological Reivew* 94: 148–175.

Kraepelin, E. 1910. *Psychiatrie*, vol. 2, 8th edition. Leipzig: Barth.

Kramer, S., Bryan, K. L., and Frith, C. D. 1998. "Confabulation" in narrative discourse by schizophrenic patients. *International Journal of Language and Communication Disorders, Suppl.* 33: 202–207.

Kripke, S. A. 1977. Identity and necessity. In *Naming, Necessity, and Natural Kinds*. Ithaca, N.Y.: Cornell University Press.

LeDoux, J. E. 1995. Emotion: Clues from the brain. *Annual Review of Psychology* 46: 209–235.

LeDoux, J. E. 2002. *Synaptic Self: How Our Brains Become Who We Are*. New York: Viking.

Lee, G. P., Arena, J. G., Meador, K. J., Smith, J. R., Loring, D. W., and Flanagan, H. F. 1988. Changes in autonomic responsiveness following amygdalectomy in humans. *Neuropsychiatry, Neuropsychology, and Behavioral Neurology* 1: 119–130.

Lewine, J. D., Andrews, R., Chez, M., Patil, A., Devinsky, O., et al. 1999. Magnetoencephalographic patterns of epileptiform activity in children with regressive autism spectrum disorders. *Pediatrics* 104: 405–418.

Lewis, S. 1987. Brain imaging in a case of Capgras' syndrome. *British Journal of Psychiatry* 150: 117–121.

Libet, B. 1996. Neural processes in the production of conscious experience. In *The Science of Consciousness*, ed. M. Velmans. London: Routledge.

Lincoln, A. 1846/1953. Letter to Allen N. Ford. In *The Collected Works of Abraham Lincoln*, ed. R. P. Basler. New York: DaCapo Press, Inc.

Lindsay, D. S., Johnson, M. K., and Kwon, P. 1991. Developmental changes in memory source monitoring. *Journal of Experimental Child Psychology* 52: 297–318.

Llinás, R. 2001. *I of the Vortex: From Neurons to Self*. Cambridge, Mass.: MIT Press.

Loftus, E. F. and Pickrell, J. E. 1995. The formation of false memories. *Psychiatric Annual* 25: 720–725.

Logue, V., Durward, M., Pratt, R. T. C., Piercy, M., and Nixon, W. L. B. 1968. The quality of survival after rupture of an anterior cerebral aneurysm. *British Journal of Psychiatry* 114: 137–160.

Losonsky, M. 1997. Self-deceivers' intentions and possessions. *Behavioral and Brain Sciences* 20: 121–122.

Lu, L. H., Barrett, A. M., Schwartz, R. L., Cibula, J. E., Gilmore, R. L., Uthman, B. M., and Heilman, K. M. 1997. Anosognosia and confabulation during the Wada test. *Neurology* 49: 1316–1322.

Luria, A. R. 1966. *Higher Cortical Functions in Man*. New York: Basic Books.

Luria, A. R. 1978. The human brain and conscious activity. In *Consciousness and Self-Regulation*, vol. 2, eds. G. E. Schwartz and D. Shapiro. New York: Plenum.

Lykken, D. T. 1957. A study of anxiety and the sociopathic personality. *Journal of Abnormal Social Psychology* 55: 6–10.

MacLean, P. D. 1954. The limbic system and its hippocampal formation. *Journal of Neurosurgery* 11: 29–44.

Mangina, C. A. and Beuzeron-Mangina, J. H. 1996. Direct electrical stimulation of specific human brain structures and bilateral electrodermal activity. *International Journal of Psychophysiology* 22: 1–8.

Mangun, G. R., Luck, S. J., Plager, R., Loftus, W., Hillyard, S., Handy, T., Clark, V. P., and Gazzaniga, M. S. 1994. Monitoring the visual world: Hemispheric asymmetries and subcortical processes in attention. *Journal of Cognitive Neuroscience* 6: 267–275.

Mather, M., Johnson, M. K., and DeLeonardis, D. M. 1999. Stereotype reliance in source monitoring: Age differences and neuropsychological test correlates. *Cognitive Neuropsychology* 16: 437–458.

Mathis, H. 1970. "Emotional response in the antisocial personality." Ph.D. dissertation, George Washington University, Washington, D.C. Available from University Microfilms, Ann Arbor, Mich., no. 71–12, 299.

Mattioli, F., Miozzo, A., and Vignolo, L. A. 1999. Confabulation and delusional misidentification: A four-year follow-up study. *Cortex* 35: 413–422.

Mavaddat, N., Kirkpatrick, P. J., Rogers, R. D., and Sahakian, B. J. 2000. Deficits in decision-making in patients with aneurysms of the anterior communicating artery. *Brain* 123: 2109–2117.

McCarthy, G., Puce, A., Belger, A., and Allison, T. 1999. Electrophysiological studies of human face perception II: Response properties of face-specific potentials generated in occipitotemporal cortex. *Cerebral Cortex* 9: 431–444.

McClelland, J. L., McNaughton, B. L., and O'Reilly, R. C. 1995. Why are there no complementary learning systems in the hippocampus and neocortex: Insights from the successes and failures of connectionist models of learning and memory. *Psychological Review* 102: 419–457.

McDaniel, K. D. and McDaniel, L. D. 1991. Anton's syndrome in a patient with posttraumatic optic neuropathy and bifrontal contusions. *Archives of Neurology* 48: 101–105.

Mele, A. R. 1987. *Irrationality: An Essay on Akrasia, Self-Deception, and Self-Control.* Oxford: Oxford University Press.

Mele, A. R. 2001. *Self-Deception Unmasked.* Princeton, N.J.: Princeton University Press.

Mendez, M. F. 1992. Delusional misidentification of persons in dementia. *British Journal of Psychiatry* 160: 414–416.

Mentis, M. J., Weinstein, E. A., Horwitz, B., McIntosh, A. R., Pietrini, P., Alexander, G. E., Furey, M., and Murphy, D. G. 1995. Abnormal brain glucose metabolism in the delusional misidentification syndromes: A positron emission tomography study in Alzheimer disease. *Biological Psychiatry* 38: 438–449.

Mercer, B., Wapner, W., Gardner, H., and Benson, D. F. 1977. A study of confabulation. *Archives of Neurology* 34: 429–433.

Mesulam, M., Van Hoesen, G. W., Pandya, D. N., and Geschwind, N. 1977. Limbic and sensory connections of the inferior parietal lobule (area pg) in the rhesus monkey: A study with a new method for horseradish peroxidase histochemistry. *Brain Research* 136: 393–414.

Mesulam, M. M. 1981. A cortical network for directed attention and unilateral neglect. *Annals of Neurology* 10: 309–325.

Metcalfe, J., Funnell, M., and Gazzaniga, M. S. 1995. Right hemisphere memory superiority: Studies of a split-brain patient. *Psychological Science* 6: 157–164.

Miller, M. B. and Gazzaniga, M. S. 1998. Creating false memories for visual scenes. *Neuropsychologia* 36: 513–520.

Miller, L. A., Caine, D., Harding, A., Thompson, E. J., Large, M., and Watson, J. D. G. 2001. Right medial thalamic lesion causes isolated retrograde amnesia. *Neuropsychologica* 39: 1037–1046.

Mindus, P., Rasmussen, S. A., Lindquist, C., and Noren, G. 2001. Neurological treatment for refractory obsessive-compulsive disorder. In *The Frontal Lobes and Neuropsychiatric Illness.* eds. S. P. Salloway, P. F. Malloy, and J. D. Duffy. Washington, D.C.: American Psychiatric Publishing.

Mishkin, M. 1978. Memory in monkeys severely impaired by combined but not by separate removal of amygdala and hippocampus. *Nature: London* 273: 297–298.

Mishkin, M., Ungerleider, L. G., and Macko, K. A. 1983. Object vision and spatial vision: Two cortical pathways. *Trends in Neuroscience* 6: 414–417.

Moore, G. E. 1942. *The Philosophy of G.E. Moore,* ed. P. A. Schilpp. Evanston, Ill.: Northwestern University Press.

Morecraft, R. J., Geula, C., and Mesulam, M. M. 1992. Cytoarchitecture and neural afferents of orbitofrontal cortex in the brain of the monkey. *Journal of Comparative Neurology* 323: 341–358.

Morecraft, R. J., Geula, C., and Mesulam, M. M. 1993. Architecture of connectivity within a cingulo-fronto-parietal neurocognitive network for directed attention. *Archives of Neurology* 50: 279–284.

Morris, J. S., Ohman, A., and Dolan, R. J. 1998. Conscious and unconscious emotional learning in the human amygdala. *Nature* 393: 467–470.

Moscovitch, M. 1989. Confabulation and the frontal lobe system. In *Varieties of Memory and Consciousness: Essays in Honor of Endel Tulving*. Hillsdale, N.J.: Lawrence Erlbaum.

Moscovitch, M. 1995. Confabulation. In *Memory Distortions*, ed. D. L. Schacter. Cambridge, Mass.: Harvard University Press.

Moscovitch, M. and Melo, B. 1997. Strategic retrieval and the frontal lobes: Evidence from confabulation and amnesia. *Neuropsychologia* 35: 1017–1034.

Myslobodsky, M. S. and Hicks, L. H. 1994. The rationale for confabulation: What does it take to fabricate a story? *Neurology, Psychiatry, and Brain Research* 2: 221–228.

Nakamura, K., Kawashima, R., Ito, K., Sugiura, M., Kato, T., Nakamura, A., Hatano, K., Nagumo, S., Kubota, K., Fukuda, H., and Kojima, S. 1999. Activation of the right inferior frontal cortex during assessment of facial emotion. *Journal of Neurophysiology* 82: 1610–1614.

Narumoto, J., Okada, T., Sadato, N., Fukui, K., and Yonekura, Y. 2001. Attention to emotion modulates fMRI activity in human right superior temporal sulcus. *Cognitive Brain Research* 12: 225–231.

Nathaniel-James, D. A. and Frith, C. D. 1996. Confabulation in schizophrenia: Evidence of a new form? *Psychological Medicine* 26: 391–399.

Nathanson, M., Bergman, P. S., and Gordon, G. G. 1952. Denial of illness. *Archives of Neurology and Psychiatry* 68: 380–387.

Neafsey, E. J. 1990. Prefrontal cortical control of the autonomic nervous system: Anatomical and physiological observations. *Progress in Brain Research* 85: 147–166.

Nedjam, Z., Dalla Barba, G., and Pillon, B. 2000. Confabulation in a patient with fronto-temporal dementia and a patient with Alzheimer's disease. *Cortex* 36: 561–577.

Neziroglu, F., McKay, D., and Yaryua-Tobias, J. A. 2000. Overlapping and distinctive features of hypochondriasis and obsessive-compulsive disorder. *Journal of Anxiety Disorders* 14: 603–614.

Nieuwenhuys, R., Voogd, J., and van Huijzen, C. 1988. *The Human Central Nervous System: A Synopsis and Atlas* (3rd ed.). Berlin: Springer-Verlag.

Nisbett, R. E. and Wilson, T. D. 1977. Telling more than we can know: Verbal reports on mental processes. *Psychological Review* 84: 231–259.

Nyberg, L., Cabeza, R., and Tulving, E. 1996. PET studies of encoding and retrieval: The HERA model. *Psychonomic Bulletin Review* 3: 135–148.

Nyberg, L. 1998. Mapping episodic memory. *Behavioral Brain Research* 90: 107–114.

O'Dwyer, J. M. 1990. Co-existence of Capgras' and deClerambault's syndromes. *British Journal of Psychiatry* 156: 575–577.

OED. 1971. *Oxford English Dictionary*. Oxford: Oxford University Press.

Ohman, A., Esteves, F., Flykt, A., and Soares, J. J. F. 1993. Gateways to consciousness: Emotion, attention, and electrodermal activity. In *Progress in Electrodermal Research*, eds. J. C. Roy, W. Boucsein, D. C. Fowles, and J. H. Gruzelier. New York: Plenum.

Ongur, D., An, X., and Price, J. L. 1998. Prefrontal cortical projections to the hypothalamus in macaque monkeys. *Journal of Comparative Neurology* 401: 480–505.

Ongur, D. and Price, J. L. 2000. The organization of networks within the orbital and medial prefrontal cortex of rats, monkeys, and humans. *Cerebral Cortex* 10: 206–219.

O'Reilly, R. and Malhotra, L. 1987. Capgras' syndrome—an unusual case and discussion of psychodynamic factors. *British Journal of Psychiatry* 151: 263–265.

O'Riordan, M. A., Plaisted, K. C., Driver, J., and Baron-Cohen, S. 2001. Superior visual search suggests higher degree of attention in the periphery. *Journal of Experimental Psychology: Human Perception and Performance* 27: 719–730.

Paillere-Martinot, M. L., Dao-Castallana, M. H., Masure, M. C., Pillon, B., and Martinot, J. L. 1994. Delusional misidentification: A clinical, neuropsychological and brain imaging case study. *Psychopathology* 27: 200–210.

Pallis, C. A. 1955. Impaired identification of faces and places with agnosia for colors: Report of a case due to cerebral embolism. *Journal of Neurology, Neurosurgery and Psychiatry* 18: 218–224.

Pandya, D. N. and Seltzer, B. 1986. The topography of commissural fibers. In *Two Hemispheres—One Brain: Functions of the Corpus Callosum*, eds. F. Lepore, M. Ptito, and H. H. Jasper. New York: Alan R. Liss.

Pandya, D. N. and Yeterian, E. H. 2001. The anatomical substrates of emotional behavior: The role of the cerebral cortex. In *Handbook of Neuropsychology* (2nd ed.). vol. 5, ed. G. Gainotti. New York: Elsevier.

Panskepp, J. 1998. *Affective Neuroscience: The Foundations of Human and Animal Emotions*. Oxford: Oxford University Press.

Papez, J. 1937. A proposed mechanism for emotion. *Archives of Neurology and Psychiatry* 38: 725–743.

Parkin, A. J. 1984. Amnesic syndrome: A lesion-specific disorder. *Cortex* 20: 743–752.

Parkin, A. J. and Leng, N. R. C. 1993. *Neuropsychology of the Amnesic Syndrome*. Hove, U.K.: Lawrence Erlbaum.

Pears, D. 1984. *Motivated Irrationality*. Oxford: Oxford University Press.

Penfield, W. 1975. *The Mystery of the Mind*. Princeton, N.J.: Princeton University Press.

Pennebaker, J. W. and Chew, C. H. 1985. Behavioral inhibition and electrodermal activity during deception. *Journal of Personality and Social Psychology* 49: 1427–1433.

Perner, J. and Wimmer, H. 1985. "John thinks that Mary thinks that....": Attribution of second-order false beliefs by 5- to 10-year old children. *Journal of Experimental Child Psychology* 39: 437–471.

Perrett, D. I., Hietanen, J. K., Oram, M. W., and Benson, P. J. 1992. Organization and function of cells responsive to faces in the temporal cortex. *Philosophical Transactions of the Royal Society of London, Series B* 335: 23–30.

Petrides, M. and Iversen, S. 1979. Restricted posterior parietal lesions in the rhesus monkey and performance on visuospatial tasks. *Brain Research* 161: 63–77.

Phelps, E. A. and Gazzaniga, M. S. 1992. Hemispheric differences in mnemonic processing: The effects of left hemisphere interpretation. *Neuropsychologia* 30: 293–297.

Pick, A. 1903. Clinical studies III: On reduplicative paramnesia. *Brain* 26: 260–267.

Pick, A. 1905. Zür Psychologie der Confabulation. *Neurologische Centralblatt* 24: 509–516.

Pick, A. 1915. Beitrag zür Psychologie des denkverlaufes beim Korsakow. *Zeitshrift für die Gesante Neurologie und Psychiatrie* 28: 344–383.

Pickup, G. J. and Frith, C. D. 2001. Theory of mind impairments in schizophrenia: Symptomatology, severity, and specificity. *Psychological Medicine* 31: 207–220.

Pierce, K., Muller, R. A., Ambrose, J., Allen, G., and Courchesne, E. 2001. Face processing occurs outside the fusiform 'face area' in autism: Evidence from functional MRI. *Brain* 124: 2059–2073.

Pisani, A., Marra, C., and Silveri, M. C. 2000. Anatomical and psychological mechanism of reduplicative misidentification syndromes. *Neurological Science* 21: 324–328.

Pollock, J. L. 1986. *Contemporary Theories of Knowledge.* Totowa, N.J.: Rowan and Littlefield.

Posner, M. and Raichle, M. 1994. *Images of Mind.* New York: Scientific American Library.

Premack, D. and Woodruff, G. 1978. Does the chimpanzee have a theory of mind? *Behavioral and Brain Sciences* 1: 515–526.

Prigatano, G. P. and Schacter, D. L. 1991. Editors' introduction to *Awareness of Deficit after Brain Injury: Clinical and Theoretical Issues.* Oxford: Oxford University Press.

Ptak, R., Birtoli, B., Imbeden, H., Hauser, C., Weis, J., and Schnider, A. 2001. Hypothalamic amnesia with spontaneous confabulations: A clinicopathologic study. *Neurology* 56: 1597–1600.

Putnam, H. 1971. It ain't necessarily so. In *Readings in the Philosophy of Language*, eds. J. F. Rosenberg and C. Travis. Englewood Cliffs, N.J.: Prentice-Hall.

Quine, W. V. O. 1969. Epistemology naturalized. In *Ontological Relativity and Other Essays.* New York: Columbia University Press.

Raine, A., Reynolds, G. P., and Sheard, C. 1991. Neuroanatomical correlates of skin conductance orienting in normal humans: A magnetic resonance imaging study. *Psychophysiology* 28: 548–558.

Raine, A., Lencz, T., Bihrle, S., LaCasse, L., and Colletti, P. 2000. Reduced prefrontal grey matter volume and reduced autonomic activity in antisocial personality disorder. *Archives of General Psychiatry* 57: 119–127.

Ramachandran, V. S. 1995. Anosognosia in parietal lobe syndrome. *Consciousness and Cognition* 4: 22–51.

Ramachandran, V. S. 1996a. The evolutionary biology of self-deception, laughter, dreaming, and depression: Some clues from anosognosia. *Medical Hypotheses* 47: 347–362.

Ramachandran, V. S. 1996b. What neurological syndromes can tell us about human nature: Some lessons from phantom limbs, Capgras syndrome, and anosognosia. *Cold Spring Harbor Symposia on Quantitative Biology* 65: 115–134.

Ramachandran, V. S. and Blakeslee, S. 1998. *Phantoms in the Brain: Probing the Mysteries of the Human Mind.* New York: William Morrow.

Ramachandran, V. S. and Churchland, P. S. 1994. Filling-in: Why Dennett is wrong. In *Dennett and His Critics*, ed. Bo Dahlbom. Oxford: Blackwell.

Ramachandran, V. S. and Hirstein, W. 1998. The perception of phantom limbs: The D.O. Hebb lecture. *Brain* 121: 1603–1630.

Ramachandran, V. S. and Rogers-Ramachandran, D. 1996. Denial of disabilities in anosognosia. *Nature* 382: 501.

Ramachandran, V. S., Altschuler, E. L., and Hillyer, S. 1997. Mirror agnosia. *Proceedings of the Royal Society of London, Series B* 264: 645–647.

Ray, J. P. and Price, J. L. 1993. Organization of projections from the mediodorsal nucleus of the thalamus to orbital and medial prefrontal cortex in macaque monkeys. *Journal of Comparative Neurology* 337: 1–31.

Redlich, F. C. and Bonvicini, G. 1908. Über das Fehlen der Wahrnehmung der eigenen Blindheit bei Hirnkrankheiten. *Jahrbuch für Psychiatry* 29: 1–134.

Rempel-Clower, N. L. and Barbas, H. 1998. Topographic organization of connections between the hypothalamus and prefrontal cortex in the rhesus monkey. *Journal of Comparative Neurology* 398: 393–419.

Ribot, T. 1881. *Les Maladies de la Mémoire*. Paris: J.B. Balliere. Trans. J. Fitzgerald as *The Diseases of Memory*. New York: Humboldt, 1883.

Richardson, E. D. and Malloy, P. F. 2001. The frontal lobes and content-specific delusions. In *The Frontal Lobes and Neuropsychiatric Illness*. eds. S. P. Salloway, P. F. Malloy, and J. D. Duffy. Washington, D.C.: American Psychiatric Publishing.

Rizzolatti, G. and Berti, A. 1990. Neglect as a neural representation deficit. *Revue Neurologique (Paris)* 146: 626–634.

Rizzolati, G., Berti, A., and Gallese, V. 2000. Spatial neglect: Neurophysiological bases, cortical circuits and theories. In *Handbook of Neuropsychology*, eds. T. Boller, J. Grafman, and G. Rizzolatti. New York: Elsevier.

Rizzolatti, G., Fadiga, L., Gallese, V., and Fogassi, L. 1996. Premotor cortex and the recognition of motor actions. *Cognitive Brain Research* 3: 131–141.

Rolls, E. T. 1999. *The Brain and Emotion*. Oxford: Oxford University Press.

Rolls, E. T., Hornak, J., Wade, D., and McGrath, J. 1994. Emotion-related learning in patients with social and emotional changes associated with frontal lobe damage. *Journal of Neurology, Neurosurgery, and Psychiatry* 57: 1518–1524.

Rylander, G. 1948. Personality analysis before and after frontal lobotomy. In *The Frontal Lobes: Proceedings of the Association for Research in Nervous and Mental Disease*, vol. 27, eds. J. F. Fulton, C. D. Aring, and S. B. Wortis. Baltimore: Williams and Wilkins.

Sackheim, H. and Gur, R. 1985. Voice recognition and the ontological status of self-deception. *Journal of Personality and Social Psychology* 48: 1365–1368.

Sacks, O. 1998. *The Man Who Mistook His Wife for a Hat*. New York: Touchstone Press.

Saleem, K. S., Suzuki, S., Tanaka, K., and Hashikawa, T. 2000. Connections between anterior inferotemporal cortex and superior temporal sulcus regions in the macaque monkey. *Journal of Neuroscience* 20: 5083–5101.

Sandifer, P. H. 1946. Anosognosia and disorders of the body schema. *Brain* 69: 122–137.

Sanides, F. 1969. Comparative architectonics of the neocortex of mammals and their evolutionary interpretation. *Annals of the New York Academy of Sciences* 167: 404–423.

Saxena, S., Brody, A. L., Schwartz, J. M., and Baxter, L. R. 1998. Neuroimaging and frontal-subcortical circuitry in obsessive-compulsive disorder. *British Journal of Psychiatry, Suppl.* 35: 26–37.

Schacter, D. L. 1990. Toward a cognitive neuropsychology of awareness: Implicit knowledge and anosognosia. *Journal of Clinical and Experimental Neuropsychology* 12: 155–178.

Schacter, D. L., Harbluk, J. L., and McLachlan, D. R. 1984. Retrieval without recollection: An experimental analysis of source amnesia. *Journal of Verbal Learning and Verbal Behavior* 23: 593–611.

Schacter, D. L., Norman, K. A., and Koutstaal, W. 1998. The cognitive neuroscience of constructive memory. *Annual Review of Psychology* 49: 289–318.

Schacter, D. L., Reiman, E., Curran, T., Yun, L. S., Bandy, D., McDermott, K. B., and Roediger, H. L. 1996a. Neuroanatomical correlates of veridical and illusory recognition memory: Evidence from positron emission tomography. *Neuron* 17: 267–274.

Schacter, D. L., Verfaellie, M., and Pradere, D. 1996b. The neuropsychology of memory illusions: False recall and recognition in amnesic patients. *Journal of Memory and Language* 35: 319–334.

Schmitt, F. F. 1988. Epistemic dimensions of deception. In *Perspectives on Self-Deception*, eds. B. P. McLaughlin and A. O. Rorty. Berkeley: University of California Press.

Schnider, A. 2001. Spontaneous confabulation, reality monitoring, and the limbic system: A review. *Brain Research Reviews* 36: 150–160.

Schnider, A., Treyer, V., and Buck, A. 2000. Selection of currently relevant memories by the human posterior medial orbitofrontal cortex. *Journal of Neuroscience* 20: 5880–5884.

Schnider, A., von Daniken, C., and Gutbrod, K. 1996. The mechanisms of spontaneous and provoked confabulation. *Brain* 119: 1365–1375.

Schwartz, J. M. 1998. Neuroanatomical aspects of cognitive-behavioral theory response in obsessive-compulsive disorder. An evolving perspective on brain and behavior. *British Journal of Psychiatry, Suppl.* 35: 38–44.

Scoville, W. B. and Milner, B. 1957. Loss of recent memory after bilateral hippocampal lesions. *Journal of Neurology, Neurosurgery, and Psychiatry* 20: 11–21.

Searle, J. R. 1983. *Intentionality: An Essay in the Philosophy of Mind.* Cambridge: Cambridge University Press.

Seneca, L. c55/1991. *Epistulae Morales, Liber V,* L2. Quoted in and trans. E. Bisiach and G. Geminiani, Anosognosia related to hemiplegia and hemianopia. In *Awareness of Deficit after Brain Injury: Clinical and Theoretical Issues,* eds. G. P. Prigatano and D. L. Schacter. Oxford: Oxford University Press.

Sequeira, H. and Roy, J. C. 1993. Cortical and hypothalamo-limbic control of electrodermal responses. In *Progress in Electrodermal Research,* eds. J. C. Roy, W. Boucsein, D. C. Fowles, and J. H. Gruzelier. New York: Plenum.

Sergent, J. and Signoret, J. L. 1992. Functional and anatomical decomposition of face processing: Evidence from prosopagnosia and PET study of normal subjects. *Philosophical Transactions of the Royal Society of London, Series B* 335: 55–62.

Serizawa, T., Saeki, N., and Yamaura, A. 1997. Microsurgical anatomy and clinical significance of the anterior communicating artery and its perforating branches. *Neurosurgery* 40: 1211–1218.

Shallice, T. 1988. *From Neuropsychology to Mental Structure.* Cambridge: Cambridge University Press.

Shimamura, A. P. 1995. Memory and frontal lobe function. In *The Cognitive Neurosciences.* ed. Michael S. Gazzaniga. Cambridge, Mass.: MIT Press.

Shimamura, A. P., Janowsky, J. S., and Squire, L. R. 1990. Memory for the temporal order of events in patients with frontal lobe lesions. *Neuropsychologia* 28: 801–813.

Siegal, M., Carrington, J., and Radel, M. 1996. Theory of mind and pragmatic understanding following right hemisphere damage. *Brain and Language* 53: 40–50.

Signer, S. F. 1994. Localization and lateralization in the delusion of substitution. *Psychopathology* 27: 168–176.

Signer, S. F. and Isbister S. R. 1987. Capgras' syndrome, de Clerambault's syndrome and *folie a deux. British Journal of Psychiatry* 151: 402–404.

Silva, J. A., Leong, G. B., Wine, D. B., and Saab, S. 1992. Evolving misidentification syndrome and facial recognition deficits. *Canadian Journal of Psychiatry* 37: 239–241.

Sims, A. and White, A. 1973. Coexistance of the Capgras and de Clerambault syndrome. A case history. *British Journal of Psychiatry* 123: 635–637.

Singer, W. 1999. Neuronal synchrony: A versatile code for the definition of relations? *Neuron* 24: 49–65.

Smith, M. L., Kates, M. H., and Vreizen, E. R. 1992. The development of frontal lobe functions. In *Handbook of Neuropsychology*, eds. F. Boller and J. Grafman. New York: Elsevier.

Socrates. 1973. *Apology*. In *The Collected Dialogues of Plato*, eds. E. Hamilton and H. Cairns. Princeton, N.J.: Princeton University Press.

Spinoza, B. 1677/1883. *The Ethics*. Trans. R. H. M. Elwes. London: Dover.

Starkstein, S. E., Federoff, J. P., Price, T. R., Leiguarda, R., and Robinson, R. G. 1992. Anosognosia in patients with cerebrovascular lesions. *Stroke* 23: 1446–1453.

Starkstein, S. E. and Robinson, R. G. 1997. Mechanisms of disinhibition after brain lesions. *Journal of Nervous and Mental Disease* 185: 108–114.

Staton, R. D., Brumback R. A., and Wilson H. 1982. Reduplicative paramnesia: A disconnection syndrome of memory. *Cortex* 18: 23–36.

Stengel, E. 1952. The patients' attitude to leucotomy and its effects. *Journal of Mental Science* 98: 382–388.

Stephens, G. L. and Graham, G. 2004. Reconceiving delusion. *International Review of Psychiatry* (in press).

Stich, S. 1983. *From Folk Psychology to Cognitive Science: The Case Against Belief.* Cambridge, Mass.: MIT Press.

Stone, V. E., Baron-Cohen, S., and Knight, R. T. 1998. Frontal lobe contributions to theory of mind. *Journal of Cognitive Neuroscience* 10: 640–656.

Storms, M. D. and Nisbett, R. E. 1970. Insomnia and the attribution process. *Journal of Personality and Social Psychology* 2: 319–328.

Streit, M., Ioannides, A. A., Liu, L., Wolwer, W., Dammers, J., Gross, J., Gaebel, W., and Muller-Gartner, H. W. 1999. Neurophysiological correlates of the recognition of facial expression of emotion as revealed by magnetoencephalography. *Cognition and Brain Research* 7: 481–491.

Stuss, D. T. 1991. Disturbances of self-awareness after frontal system damage. In *Awareness of Deficit after Brain Injury: Clinical and Theoretical Issues*, eds. G. P. Prigatano and D. L. Schacter. Oxford: Oxford University Press.

Stuss, D. T., Alexander, M. P., Lieberman, A., and Levine, H. 1978. An extraordinary form of confabulation. *Neurology* 28: 1166–1172.

Stuss, D. T. and Benson, D. F. 1986. *The Frontal Lobes.* New York: Raven.

Stuss, D. T., Gallup, G. G., and Alexander, M. P. 2001. The frontal lobes are necessary for "theory of mind." *Brain* 124: 279–286.

Sully, J. 1888. *Illusions: A Psychological Study.* New York: Appleton.

Swartz, B. E. and Brust, J. C. 1984. Anton's syndrome accompanying withdrawal hallucinations in a blind alcoholic. *Neurology* 34: 969–973.

Talland, G. A. 1961. Confabulation in the Wenrnicke-Korsakoff syndrome. *Journal of Nervous and Mental Disease* 132: 361–381.

Talland, G. A. 1965. *Deranged Memory.* New York: Academic Press.

Thompson, A. E. and Swan, M. 1993. Capgras' syndrome presenting with violence following heavy drinking. *British Journal of Psychiatry* 162: 692–694.

Todd, J., Dewhurst, K., and Wallis, G. 1981. The syndrome of Capgras. *British Journal of Psychiatry* 139: 319–327.

Torvik, A., Lindboe, C. F., and Rogde, S. 1982. Brain lesions in alcoholics. *Journal of Neurological Science* 56: 233–248.

Tranel, D. 2000. Electrodermal Activity in cognitive neuroscience: Neuroanatomical and neurophysiological correlates. In *Cognitive Neuroscience of Emotion*, eds. R. D. Lane and L. Nadel. Oxford: Oxford University Press.

Tranel, D. and Damasio, A. R. 1985. Knowledge without awareness: An autonomic index of facial recognition by prosopagnosics. *Science* 228: 1453–1454.

Tranel, D. and Damasio, A. R. 1986. Covert recognition of "signal" stimuli after bilateral amygdala damage. *Society for Neuroscience Abstracts* 12: 21.

Tranel, D. and Hyman, B. T. 1990. Neuropsychological correlates of bilateral amygdala damage. *Archives of Neurology* 47: 349–355.

Trevarthen, C. B. and Sperry, R. W. 1973. Perceptual unity of the ambient visual field in human commissurotomy patients. *Brain* 96: 547–570.

Treves, A. and Rolls, E. T. 1994. Computational analysis of the role of the hippocampus in memory. *Hippocampus* 4: 374–391.

Tulving, E. 1987. Multiple memory systems and consciousness. *Human Neurobiology* 6: 67–80.

Tulving, E., Kapur, S., Craik, F. I., Moscovitch, M., and Houle, S. 1994. Hemispheric encoding/retrieval asymmetry in episodic memory: Positron emission tomography findings. *Proceedings of the National Academy of Sciences U.S.A.* 91: 2016–2020.

Van Essen, D. C. and Maunsell, J. H. R. 1983. Hierarchical organization and functional streams in the visual cortex. *Trends in Neuroscience* 6: 370–375.

Van Wagenen, W. P. and Herren, R. Y. 1940. Surgical division of commissural pathways in the corpus callosum: Relation to the spread of an epileptic attack. *Archives of Neurology and Psychiatry* 44: 740–759.

Victor, M., Adams, R. D., and Collins, G. H. 1971. *The Wernicke-Korsakoff Syndrome*. Philadelphia: F. A. Davis.

Vilkki, J. 1985. Amnesic syndromes after surgery of anterior communicating artery aneurysms. *Cortex* 21: 431–444.

Vogt, B. A. and Pandya, D. N. 1987. Cingulate cortex of the rhesus monkey: II. Cortical afferents. *Journal of Comparative Neurology* 262: 271–289.

von Monakow, C. 1885. Experimentelle und pathologisch-anatomische Untersuchungen über die Beziehungen der soggenannten Sehsphare zu den infracorticalen Opticuscentren und zum N opticus. *Archives of Psychiatry* 16: 151–199.

Waid, W. W. and Orne, M. T. 1981. Reduced electrodermal response to conflict, failure to inhibit dominant behavior, and delinquent proneness. *Journal of Personality and Social Psychology* 43: 769–774.

Wallis, G. 1986. Nature of the misidentified in the Capgras' syndrome. *Bibliotheca Psychiatrica* 164: 40–48.

Watson, R. T., Miller, B. D., and Heilman, K. M. Nonsenory neglect. 1978. *Annals of Neurology* 3: 505–508.

Wegner, D. M. 2002. *The Illusion of Conscious Will*. Cambridge, Mass.: MIT Press.

Weinstein, E. A. 1971. Linguistic aspects of amnesia and confabulation. *Journal of Psychiatric Research* 8: 439–444.

Weinstein, E. A. 1996. Symbolic aspects of confabulation following brain injury: Influence of premorbid personality. *Bulletin of the Menninger Clinic* 60: 331–350.

Weinstein, E. A. and Kahn, R. L. 1955. *Denial of Illness: Symbolic and Physiological Aspects*. Springfield, Ill.: Charles C Thomas.

Weinstein, E. A., Kahn, R. L., and Malitz, S. 1956. Confabulation as a social process. *Psychiatry* 19: 383–396.

Weiskrantz, L. 1986. *Blindsight: A Case Study and Implications*. Oxford: Oxford University Press.

Wells, C. E. and Whitehouse, P. J. 1996. Cortical dementia. In *Neuropsychiatry*, eds. B. S. Fogel, R. B. Schiffer, and S. M. Rao. Baltimore: Williams and Wilkins.

Welt, L. 1888. Über Charakterveränderungen des Menschen infolge von Läsionen des Stirnhirns. *Deutsches Archiv für Klinische Medicin* 42: 339–390.

Wernicke, C. 1906. *Grundriss der Psychiatrie* (2nd ed.). Leipzig: Thieme.

Whalen, P. J., Rauch, S. L., Etcoff, N. L., McInerney, S. C., Lee, M. B., and Jenike, M. A. 1998. Masked presentation of emotional facial expressions modulate amygdala activity without explicit knowledge. *Journal of Neuroscience* 18: 411–418.

Whitlock, F. A. 1981. Some observations on the meaning of confabulation. *British Journal of Medical Psychology* 54: 213–218.

Whitty, C. W. and Lewin, W. 1957. Vivid day-dreaming: An unusual form of confusion following anterior cingulectomy in man. *Brain* 80: 72–76.

Winner, E., Brownell, H., Happe, F., Blum, A., and Pincus, D. 1998. Distinguishing lies from jokes: Theory of mind deficits and discourse interpretation in right hemisphere brain-damaged patients. *Brain and Language* 62: 89–106.

Wittgenstein, L. 1969. *On Certainty*. Oxford: Blackwell.

Wright, S., Young, A. W., and Hellawell, D. J. 1993. Sequential Cotard and Capgras delusions. *British Journal of Clinical Psychology* 32: 345–349.

Young, A. W. 1998. *Face and Mind*. Oxford: Oxford University Press.

Young, A. W. 2000. Wondrous strange: The neuropsychology of abnormal beliefs. In *Pathologies of Belief*, eds. M. Coltheart and M. Davies. Oxford: Blackwell.

Young, A. W., Hellawell, D. J., Wright, S., and Ellis, H. D. 1994. Reduplication of visual stimuli. *Behavioral Neurology* 7: 135–142.

Zaidel, E. 1976. Auditory vocabulary of the right hemisphere following brain bisection or hemidecortication. *Cortex* 12: 191–211.

Zaidel, E. 1983. A response to Gazzaniga: Language in the right hemisphere, convergent perspectives. *American Psychologist* 38: 542–546.

Zald, D. H. and Kim, S. W. 2001. The orbitofrontal cortex. In *The Frontal Lobes and Neuropsychiatric Illness*. eds. S. P. Salloway, P. F. Malloy, and J. D. Duffy. Washington, D.C.: American Psychiatric Publishing.

Zangwill, O. L. 1953. Disorientation for age. *Journal of Mental Science* 99: 698–701.

Zilbovicius, M., Boddaert, N., Pascal, B., Poline, J., Remy, P., Mangin, J., Thivard, L., Barthelemy, C., and Samson, Y. 2000. Temporal lobe dysfunction in childhood autism: A PET study. *American Journal of Psychiatry* 157: 1988–1993.

Zoccolatti, P., Scabini, D., and Violani, C. 1982. Electrodermal responses in unilateral brain damage. *Journal of Clinical Neuropsychology* 4: 143–150.

Name Index

Ackil, J. K., 13, 67
Adair, J. C., 140, 142
Adolphs, R., 110–111, 150
Albert, M. L., 78, 135, 146
Alexander, M. P., 118, 125–126
Amaral, D. G., 97
Anderson, M. W., 235
Anderson, S. W., 92–93
Anton, G., 12, 136, 140, 188–189
Armstrong, D. M., 39

Babinski, J., 135
Bachevalier, J., 58, 109
Bacon, A. L., 104
Bagshaw, M. W., 97
Bailey, P., 137
Barbas, H., 58, 82, 85–86, 186
Barbizet, J., 191
Baron-Cohen, S., 103, 106, 109, 112
Bauer, R. M., 22, 118–119, 122, 126
Bear, D. M., 119
Beats, B., 117–118
Bechara, A., 79, 92, 166
Beeckmans, K., 56–57, 79, 95, 167
Beers, C. W., 124
Behrman, M., 130
Bender, L., 18, 200
Benson, D. F., 3, 12, 32, 54, 61, 71,
   76, 90, 124–125, 137–138, 144–
   146, 175, 178, 191
Bergson, H., 131
Berlyne, N., 7–9, 17, 22, 51, 196, 198,
   201
Bermudez, J. L., 230
Berrios, G. E., 2, 7, 19, 199
Bertilson, H. S., 92
Berti, A., 140
Beukelman, D. R., 167
Beuzeron-Mangina, J. H., 97
Bianchi, L., 77
Bichat, F., 153
Bick, P. A., 125
Bisiach, E., 10, 136–139, 141, 143,
   147–148, 161, 171, 183, 189

Blair, R. J. R., 92
Blakeslee, S., 6, 138
Blanke, O., 151
Bleuler, E. P., 212
Blumer, D., 76, 90
Bogen, J. E., 153
Bonhoeffer, K., 2, 7, 50, 103, 202
Bonnet, C., 117–118
Bonvicini, G., 146
Box, O., 210
Breen, N., 122, 124
Breiter, H. C., 121
Brion, S., 54
Brothers, L., 105–106, 109
Brust, J. C., 12, 146
Budson, A. E., 119
Burgess, P. W., 3, 44
Burns, A., 125
Butler, P. V., 126

Cabeza, R., 64, 163
Cameron, N., 212
Campbell, R., 121
Capgras, J., 12, 126
Cappa, S. F., 138
Carey, D. P., 110
Carmichael, S. T., 82–83, 85–89, 95,
   110–111, 121
Carpenter, M. B., 55
Carter, C. S., 59
Case, R., 34, 67
Cavada, C., 168
Ceci, S. J., 67
Chao, L. L., 130
Charcot, J. M., 25
Chatterjee, A., 18
Chew, C. H., 94–95
Christodoulou, G. N., 126, 128, 188
Churchland, P. M., 38–41
Churchland, P. S., 6, 29, 38, 41, 193
Cicerone, K. D., 7
Cohen, N. J., 46
Cook, N. D., 157
Cotard, J., 116–117

Coughlan, A. K., 8, 44, 65
Courbon, P., 116–117
Critchley, H. D., 95, 144
Critchley, M., 137, 143
Cummings, J. L., 143, 146
Cunningham, J. M., 66
Curran, T., 90
Cutting, J., 50, 129

Dalla Barba, G., 62, 64, 165, 191, 209
Damasio, A. R., 22, 27, 57, 59, 73, 76, 78–80, 91–92, 97, 105, 166, 170, 174, 180, 191
Davidson, D., 188, 213–214, 217, 221–223, 225, 227, 232
Davidson, R. A., 166
Deese, J., 67
Deglin, V. L., 169–170
deHaan, E. H. F., 122
Delbecq-Derousne, J., 68
Delgado, J. M. R., 170–171
deLuca, J., 7, 9, 16, 58, 62, 66, 200
Demeter, S., 158
Dengerink, H. A., 92
Dennett, D., 5, 14, 20, 163, 241
dePauw, K. W., 114, 116, 128
Descartes, R., 25
Devinsky, O., 37, 59, 87, 173
Diamond, B. J., 9
diVirgilio, J., 158
Drake, M. E., 125
Dywan, J., 13, 68

Edelstyn, N. M., 117, 125, 128
Ehrt, U., 130
Ekman, P., 93
Elliot, R., 95
Ellis, H. D., 22, 116, 118–120, 123–124, 126, 129, 131
Eslinger, P. J., 108
Euthyphro, 195

Fail, G., 116
Farah, M. J., 184
Feinberg, T. E., 58, 128, 137, 143, 165, 188, 198, 200, 202

Fingarette, H., 216, 227–228
Finger, S., 49, 74, 77, 86
Fischer, R. S., 7, 9, 44, 57, 90, 169
Flanagan, O., 219, 242
Fleminger, S., 125
Förstl, H., 117–118, 125
Fowles, D. C., 95
Frazer, S. J., 115, 123, 130
Freedman, A. M., 16
Freeman, W., 77–78, 81
Freud, S., 71, 217
Fried, I., 120
Frith, C. D., 13, 17, 122
Fukutake, T., 80
Fulford, K. W. M., 19
Fulton, J. F., 137
Funnell, M. G., 153, 158
Fuster, J. M., 32, 48

Gaffan, D., 52, 55
Gage, P., 74, 76, 80, 239
Gainotti, G., 16, 137, 141, 157, 167–168
Gamper, E., 54
Gardner, H., 202
Gassell, M. M., 186
Gazzaniga, M. S., 10, 42, 124, 153–154, 156–157, 162–166, 169, 172, 176, 183, 210, 241
Geminiani, G., 10, 136–139, 141, 147–148, 161, 171, 183, 189
Geschwind, N., 25, 42, 137, 142, 144, 148, 152, 154, 160–162, 184–185, 188, 211
Gianetti, E., 213, 230
Gilovich, T., 213
Gloor, P., 52, 110, 160
Goddard, G. V., 109
Goktepe, E. O., 99
Goldman, A., 20, 179, 207
Goldman-Rakic, P. S., 36, 49, 88–89, 110
Grace, G. M., 129
Graham, G., 18
Grattan, L. M., 108
Greenblatt, M., 78
Green, C., 235

Gregory, C., 106
Gur, R., 229

Halgren, E., 97, 119, 235
Harding, A., 55
Hare, R. D., 91–92, 94, 102–103
Harlow, J. M., 76
Hartmann, J. A., 22
Harwood, D. G., 116
Hasselmo, M., 121
Haxby, J. V., 110, 120–122
Hecaen, H., 78, 146, 170
Heilman, K. M., 10, 94, 139–140,
  144, 146, 149–151, 171, 203, 211
Hellige, J. B., 155
Hellman, L., 234
Herren, R. Y., 153
Hicks, L. H., 200–201
Hirstein, W., 12, 22, 38, 101, 112,
  115, 119, 125–126, 139, 164, 205,
  229
Hoffman, E. A., 120–121
Holbourn, A. H. S., 124
Hopkins, M. J., 95
Hornak, J., 108
Hudson, A. J., 129
Hyman, B. T., 97

Isbister, S. R., 117
Iversen, S., 143–144

Jackson, J. H., 25, 32, 46, 72, 172,
  184
James, W., 68, 132
Jernigan, T. L., 54
Jeste, D. V., 13
Jha, A. P., 156
Johnson, C. N., 132
Johnson, M. K., 3, 9, 45, 58, 61–63,
  65, 67, 71–72, 91, 157, 168–169,
  175, 178, 211
Jolly, F., 50
Jones, B., 80
Jones, H. E., 94
Joseph, A. B., 124
Joseph, R., 137, 154, 161–162, 205,
  242

Joseph, R. M., 132
Joslyn, D., 18

Kaada, B. R., 86
Kahn, R. L., 137, 140, 152, 195
Kapur, N., 8, 44, 65, 125
Kartsounis, L. D., 99
Kern, R. S., 13, 18
Kim, S. W., 78, 85–86, 89–90, 97–99,
  108, 175
Kinsbourne, M., 139, 149, 157, 169–
  170, 241
Kolakowski, L., 131
Kopelman, M. D., 7–8, 13, 64–65,
  124, 196
Korsakoff, S. S., 8, 17, 43–44, 49–50,
  60
Kosslyn, S. M., 129
Kraepelin, E., 11, 196
Kramer, S., 13
Kripke, S. A., 15

LeDoux, J. E., 10, 27, 85, 174
Lee, G. P., 97
Leng, N. R. C., 50–52, 55, 58, 167
Lettsom, J. C., 1, 118
Lewine, J. D., 109
Lewin, W., 59–60
Lewis, S., 126
Libet, B., 173
Lincoln, A., 199
Lindsay, D. S., 67
Llinás, R., 85, 174
Loftus, E. F., 43
Logue, V., 81
Losonsky, M., 215
Luria, A. R., 25, 68, 157
Lu, L. H., 18, 165
Luzzatti, C., 138
Lykken, D. T., 92

McCarthy, G., 121
McClelland, J. L., 49
McDaniel, K. D., 12, 137, 145–146
McDaniel, L. D., 12, 137, 145–146
MacLean, P. D., 25, 36
Malhotra, L., 115

Malloy, P. F., 189
Mangina, C. A., 97
Mangun, G. R., 143
Margaret, A., 212
Marinkovic, K., 235
Mather, M., 62
Mathis, H., 92
Mattioli, F., 115, 125, 130, 186
Maunsell, J. H. R., 36
Mavaddat, N., 55–57, 79
Mele, A. R., 214, 217, 221, 223–225,
    228–229, 231–232
Melo, B., 44, 199–201
Mendez, M. F., 126
Mentis, M. J., 125–126, 152
Mercer, B., 17, 44, 51, 61, 64, 165,
    191, 201, 209
Mesulam, M. M., 144, 151, 174
Metcalfe, J., 164
Miller, L. A., 52
Miller, M. B., 157
Milner, B., 46
Mindus, P., 98
Mishkin, M., 36, 55, 57–58, 80
Moniz, E., 77
Moore, G. E., 188, 205
Morecraft, R. J., 87, 109, 144
Morris, J. S., 168
Moscovitch, M., 18–19, 44, 130, 199–
    201
Murray, E., 52
Myslobodsky, M. S., 200–201

Nakamura, K., 108
Narumoto, J., 121
Nathaniel-James, D. A., 13, 17
Nathanson, M., 136
Neafsey, E. J., 95
Nedjam, Z., 64–65
Neziroglu, F., 149
Nisbett, R. E., 3, 14
Nyberg, L., 58, 157

O'Dwyer, J. M., 117
Ohman, A., 166
Ongur, D., 87, 174
O'Reilly, R. C., 115

O'Riordan, M. A., 113
Orne, M. T., 93
Oyebode, F., 117, 125

Paillere-Martinot, M. L., 116, 124
Pallis, C. A., 118
Pandya, D. N., 3, 35–36, 83, 87, 89,
    111, 151, 155, 158, 186
Panskepp, J., 103
Papez, J., 36, 86
Parkin, A. J., 44, 50–52, 55, 58
Pears, D., 217
Penfield, W., 110, 170
Pennebaker, J. W., 94–95
Perner, J., 132
Perrett, D. I., 110, 121
Petrides, M., 143–144
Phelps, E. A., 164
Pickrell, J. E., 43
Pickup, G. J., 122
Pick, A., 7, 62, 116, 136
Pierce, K., 122
Pisani, A., 129
Plato, 217
Pollock, J. L., 204
Posner, M., 59, 99
Premack, D., 103
Price, J. L., 82–83, 85–89, 95, 110–
    111, 121, 174, 228
Prigatano, G. P., 140
Ptak, R., 60, 63
Putnam, H., 15

Quine, W. V. O., 179
Quinn, M. J., 92

Raichle, M., 59, 99
Raine, A., 92, 97
Ramachandran, V. S., 6, 10–12, 15,
    22, 101, 115, 119, 125–126, 137–
    139, 143, 150, 164, 193
Raye, C. L., 169
Ray, J. P., 85, 228
Reboul-Lachaux, J., 12, 126
Redlich, F. C., 146
Rempel-Clower, N. L., 86
Ribot, T., 46

Richardson, E. D., 189
Ring, B., 109
Rizzolatti, G., 106, 111, 140, 143–144, 151, 174
Roane, D. M., 58, 128, 165, 188, 202
Roberts, J. M., 115, 123, 130
Robinson, R. G., 80–81, 113, 124, 157
Rogers-Ramachandran, D., 150
Rolls, E. T., 28, 49, 78–80, 88, 103, 108, 171–172, 181, 183
Roy, J. C., 94, 97
Rylander, G., 98

Sackheim, H., 229
Sacks, O., 27
Saleem, K. S., 122
Sandifer, F., 11
Sanides, F., 35
Saxena, S., 22, 97
Schacter, D. L., 43–44, 48, 64, 67, 94, 140, 149, 163, 241
Schmitt, F. F., 218–220
Schnider, A., 7, 60, 63–64, 66, 68, 71, 180, 216, 227
Schwartz, J. M., 22, 97
Scoville, W. B., 46
Searle, J. R., 205, 229
Sejnowski, T., 29
Seltzer, B., 3, 155, 158
Seneca, L., 135
Sequeira, H., 94, 97
Sergent, J., 118
Serizawa, T., 57
Shallice, T., 3, 44, 173
Shimamura, A. P., 48, 65, 72
Siegal, M., 112
Signer, S. F., 12, 117, 125
Signoret, J. L., 118
Silva, J. A., 123
Sims, A., 117
Singer, W., 174
Smith, M. L., 34, 67
Socrates, 177, 195, 239
Sophocles, 40
Sperry, R. W., 153, 165
Spinoza, B., 171
Squire, L. R., 46

Stanhope, N., 49
Starkstein, S. E., 80–81, 124, 150, 157
Staton, R. D., 125–126, 128–129
Stengel, E., 78
Stephens, G. L., 18
Stich, S., 40
Stone, V. E., 106, 132
Storms, M. D., 14
Streit, M., 121
Stuss, D. T., 3, 7, 32, 44, 57, 65–66, 78, 126, 137–138, 144–146, 173, 175, 178
Sully, J., 6
Swan, M., 114
Swartz, B. E., 12, 146
Szulecka, T. K., 114

Talland, G. A., 18, 50–51, 63, 65, 184, 198, 200–201, 203, 226
Thompson, A. E., 114
Todd, J., 115
Torvik, A., 50, 54
Tranel, D., 22, 97
Trevarthen, C. B., 165
Treves, A., 48
Tulving, E., 157
Tusques, J., 117

van Essen, D. C., 36
van Wagenen, W. P., 153
Victor, M., 52
Vilkki, J., 65
Vogel, P. J., 153
Vogt, B. A., 87, 111
von Monakow, C., 136

Waid, W. W., 93
Wallis, G., 114
Watson, R. T., 143
Watts, J. W., 77–78, 81
Wegner, D. M., 171, 200, 212, 241
Weinstein, E. A., 4, 17–18, 22, 101, 137, 140, 152, 195
Weiskrantz, L., 21
Wellman, H. M., 132
Wells, C. E., 13
Welt, L., 74

Wernicke, C., 7, 49–50
Whalen, P. J., 86
Whitehouse, P. J., 13
White, A., 117
Whitlock, F. A., 7, 16, 103, 202
Whitty, C. W., 59–60
Williams, D., 186
Wilson, T. D., 3, 14
Wimmer, H., 132
Winner, E., 112
Wittgenstein, L., 187, 205
Woodruff, G., 103
Wright, S., 116, 126

Yeterian, E. H., 35–36, 83, 89, 151,
    186
Young, A. W., 22, 115–116, 118, 120,
    123, 126, 129, 131, 133, 148

Zaidel, E., 156
Zald, D. H., 78, 85, 87, 89–90, 97–99,
    108, 175
Zangwill, O. L., 16, 50
Zaragoza, M. S., 13, 67
Zilbovicius, M., 112
Zoccolatti, P., 155

Subject Index

Note: Page numbers in *italics* indicate figures.

Achromotopsia, 36, 130
Acquired sociopathy, 72, 90, 98
Akinetic mutism, 59, 183
Alien hand syndrome, 167
Alzheimer's disease, 13, 18, 65, 116,
  125
Amnesia, 43–44, 48, 65–66. *See also*
  Korsakoff's amnesia; Medial
  temporal lobe amnesia
Amygdala, 37, *47(#14)*, 48, *54*, 55, 63,
  97, *107*, 158, 175, 182
  and face perception, 97, 109, 119,
  121
  and mind reading, 106, 109–110,
  119, 128
  and orbitofrontal cortex, 80–83, 85–
  86
Analog representations, 181, 205–
  206, 210, 229
Aneurysm of the anterior communicating artery, 9, 16, 79, 81–82,
  91, 167. *See also* Anterior communicating artery
  and anosognosia, 143, 146, 151
  and definition of confabulation,
  177, 183, 201
  and memory confabulations, 44–45,
  55–60, 63, 65–66, 68–69, 71
  and the misidentification
  syndromes, 101–102, 133
Anosognosia, 1, 10–11, 17, 57, 135–
  143, 150–151, 171, 223–224,
  230
  and definition of confabulation,
  177, 183, 190, 192, 202, 204, 207
  and the two hemispheres, 158, 163,
  166
Anteromedial thalamic nucleus. *See*
  Anterior thalamic nucleus
Anterior cerebral artery, 55, 58
Anterior cingulate gyrus, *34*, 37, 55,
  58–59, *96*, 128, 170, 173, 175

and the orbitofrontal syndromes, 71,
  76, 87, 95, 99
Anterior cingulectomy, 59, 62, 133,
  242
Anterior commissure, *53(#13)*, 57–58,
  125, 153, 158–160, *159(#7)*, 167
Anterior communicating artery
  (ACoA), 55, 57–58, 87, 158, 167
Anterior communicating artery
  syndrome. *See* Aneurysm of the
  anterior communicating artery
Anterior thalamic nucleus, *47(#4)*,
  52–55, *53(#5)*, 85, 88
Anton's syndrome, 11–12, 22, 136,
  145–146, 222, 231
  and definition of confabulation,
  177–178, 190, 195, 204, 207
Aphasia, 27, 183. *See also* Wernicke's
  aphasia
Asomatognosia, 11, 136, 143
Autism, 104, 106, 109, 112–113, 122,
  134, 184
Autobiographical memory, 8, 17, 43–
  44, 48–49, 163, 167, 169, 177
  confabulations about, 45, 52, 58,
  61–65, 129, 201, 205
Autonomic nervous system, 32, 37,
  60, 82–83, 86, 93–94, 166, 180,
  182, 215
Autonomic activity, 22, 79, 85–87,
  92, 141, 155, 175, 191, 193–194
  and mind reading, 102, 105, 109–
  110, 113, 128
  and face perception, 22, 121–122
  and self-deception, 223, 231, 234

Background, 205
Basal forebrain, *34*, 55, 58, 63–64, 97
Belief criterion in the definition of
  confabulation, 16, 188–190, 225
Binding, 140, 174. *See also*
  Consciousness

Blindsight, 21
Blind spot. *See* Filling in
Body representations, 106, 139–140,
    147–148, 150

Callosotomy, 153, 156, 160, 167
Capgras' syndrome, 12–13, 17, 19,
    42, 100–101, 114–118, 142, 152
and definition of confabulation,
    177, 180, 186, 188–189, 191, 194
mind-reading theory of, 122–126,
    130–133
other theories of, 125–126, 128–130
Catastrophic reaction, 16
Caudate nucleus, 35, 89–90, 99, 143,
    159(#4)
Center of narrative gravity, 5, 241–
    242
Checking processes, 62–63, 66–67,
    147–150, 175, 214, 233, 240, 243
and definition of confabulation,
    178–181, 183–186, 191, 208
Children, 132
and false memories, 13, 67
and lying, 133
and social referencing, 104–105,
    132–133
and mind reading, 132
Cingulate gyrus, *34*, 36–37, 61, 109,
    125, 143–144. *See also* Anterior
    cingulate gyrus; Posterior cingulate
    gyrus
Claiming, 76, 102, 182, 203, 225,
    232
as a criterion for confabulation,
    188–189, 200, 202–203
Clutch, 190
Cognitive science, 26–27
Commissures, 153, 155. *See also*
    Anterior commissure; Corpus
    callosum; Hippocampal
    commissure
Commisurotomy, 153, 160
Conceptual representations, 181–182,
    185, 206, 229
Confabulation
definition of, 41, 187–203

epistemic concept of, 18, 20–21,
    176, 198
linguistic concept of, 20, 198–199
mnemonic concept of, 19, 198
in normal people, 3, 13–14, 183,
    194–195
Confabulations of embarrassment, 7,
    103
Confidence, 4, 11, 92–93, 101, 193–
    194, 207, 220, 226, 240
Conscience, 91, 243
Consciousness, 29, 36, 38, 59, 86,
    119, 149–150, 230, 235
and the two hemispheres, 168, 173–
    175
Conscious awareness system (CAS),
    149
Contradictions, 9, 18, 62, 91, 243
Contradictory beliefs, 188–191, 218,
    228–229
Corpus callosum, 3, *34*, 55, *56*, 57–
    58, 153, 158, *159*, 160, 167, 169
Cotard's syndrome, 116–117, 126–
    127, 133, 194
Creative stream, 181–184, 239–241,
    243

Deception, 108, 218–220, 229–230,
    233–234. *See also* Lying
deep vs. shallow, 219
DeClerambault's syndrome, 117, 124,
    133
Degraded representations, 169, 177,
    184–185, 191, 223, 231
Degraded representation principle,
    185
Definition, types of, 187
Delusion, 17–19, 148, 199
Denial, 94, 135, 137–138, 143–144,
    150, 152, 155, 212, 214. *See also*
    Anosognosia
of blindness (*see* Anton's syndrome)
and definition of confabulation,
    185, 195, 204
of hemianopia, 136, 177
of hemiplegia, 1, 135–136, 145, 165,
    177–178

Desire criterion in definition of confabulation, 16–17, 184, 194, 231–232
Disconnection theory of anosognosia, 142, 149
Disconnection theory of confabulation, 160–163
Disinhibition, 13, 32, 42, 99, 101–104, 108, 134, 155, 157, 167, 183–184
and the orbitofrontal syndromes, 71–72, 74, 76, 78, 80–82, 99, 145
Disorientation, 51, 136, 144, 152
Don't Know test, 65, 191, 209–211
Dorsal stream, 33, 35–36, 81, 112–114, 118–119, 122, 151
Dorsolateral prefrontal cortex, 45, 57, 82, 87–89, 128, 168, 174
Doubt, 4, 22, 97, 180, 182, 205, 207
Dreams, 242

Echolalia, 114
Eliminative materialism, 40
Embarassment, 91, 94
Emotion, 29, 36–37, 141, 180, 191–192, 228, 232, 235
and anosognosia, 144–145, 150
and mind reading, 105–106, 108, 113–114, 126
and orbitofrontal cortex, 72–74, 77–81, 83, 85, 174
perceiving in faces, 120–121, 131, 150
perceiving in speech, 156
and sociopathy, 91–92
and the two hemispheres, 154, 156, 161, 163, 166–170
Empathy, 78, 85, 91–92, 100, 108, 131
Entorhinal cortex, 47(#17), 48, 158
Epilepsy, 10, 46, 86, 109–110, 113, 120, 125, 151
and the two hemispheres, 153, 165, 167–168
Episodic memory, See Autobiographical memory
Epistemic disinhibition, 81

Epistemic concept of confabulation. See Confabulation
Epistemology, 38, 179, 203–204, 207–208
Estimation, 79, 95, 99, 208, 236
Executive processes, 44–45, 57, 61, 64, 156, 180–181
External vs. internal representations, 123–124, 127–128, 131–132, 242
Eye gaze processing. See Gaze perception

Face identity perception, 120–121, 128
Face perception, 110, 117–122
False memories, 66–67, 163
False recognition, 56, 68, 90
Falsity criterion in definition of confabulation, 18, 20, 198–200
Fantastic vs. momentary confabulation, 7–8
Faux pas, 106, 132
Fiction, 200
Filling in, 6, 17, 165, 175, 188, 193
Filtering, 72, 148, 185, 241
Folk psychology, 40, 217
Fornix, 47(#3), 53(#1), 54–55, 56, 57, 160
Fregoli's syndrome, 116–117, 123–124, 127–128, 210, 242
Frontal eye fields, 140, 143–144, 148
Frontal theories of confabulation, 3, 60–65, 166–167
Functional map, 32, 82
Fusiform gyrus, 75(#9), 120–122, 130–131

Gambling task, 79, 95
Gap filling, 17–19
Gaze perception, 109, 113, 120–122
Gyrus rectus, 74, 75(#12), 82, 87, 125

Hallucinations, 61, 80, 117, 122, 124, 146, 171
Hemianopia, 135–136, 148
Hemiplegia, 135–137, 148

HERA model of memory, 157, 167
Hippocampal commissure, 153, 158–160
Hippocampus, 35–37, 47, 67, 82, 88, 109–110, 160
and memory confabulations, 46, 48–49, 55, 58
Holism vs. localizationism, 30
"Holy Ghost," 98
Homunculus fallacy, 163–164, 185
Hypnosis, 13–14, 68
Hypochondria, 194. See also Mirror-image syndromes
Hypothalamus, 36, 56, 57–58, 60, 63, 143
and the autonomic system, 37, 83, 86, 128, 169, 180
and the orbitofrontal syndromes, 77–78, 81, 87, 95, 97

Ill-grounded thoughts, 18, 20–21, 42, 102, 178–179, 190–197, 204, 241
and self-deception, 214–215, 222, 225–226
Illusion of subjective doubles, 117
Imaging, 28
Implausible vs. plausible confabulation, 195–196
Inferior parietal cortex, 58, 95–97, 96, 99, 110–111, 119, 125, 180–181, 190
and anosognosia, 143–145, 147–148, 150–152
and the two hemispheres, 161, 174
Inhibition, 63–64, 100, 119, 133, 157, 170, 176, 228, 232, 239. See also Disinhibition
and definition of confabulation, 180, 186–187, 208
and the orbitofrontal syndromes, 73, 85, 89, 94–95
Injection experiment, 138
Insula, 35–37, 87–88, 109
Integrated field theory of consciousness, 149, 174
Intentional action, 170–174

Intentional confabulation, 15–16, 194, 198
Intentional deception, 15–17, 92, 194, 198, 202
Intentional self-deception, 214–217, 229–231
Intermetamorphosis, 117, 123, 125
Internalism vs. externalism in epistemology, 204
Internal representations. See External vs. internal representations
Interpreter, 162–164, 172, 183, 234, 239
Intralaminar thalamic nuclei, 85, 87, 143, 174
Irony, 112, 221

Jokes, 106, 112

Korsakoff's amnesia, 7–9, 17, 49–50, 55
Korsakoff's syndrome, 1, 2, 7–9, 13, 101–102, 161, 167, 226
and anosognosia, 144, 146, 151
and definition of confabulation, 177, 180, 190, 192, 196, 199, 201, 207, 210
and memory confabulations, 43–45, 49–57, 62, 65–66, 69
and the misidentification syndromes, 129, 133
and the orbitofrontal syndromes, 71, 82, 85, 88, 91

Lateral prefrontal cortex. See Dorsolateral prefrontal cortex
Lateral theories of confabulation, 3, 154–155, 160–164
Laterality, 94, 112, 142–143, 155–157, 167–170
Left hemisphere, 11, 95, 140, 143, 148, 154–157, 162–170, 234
and definition of confabulation, 179, 184–185
Limbic system, 36–37, 52, 60–61, 63, 82–83, 87, 109, 161, 235

Linguistic concept of confabulation. *See* Confabulation
Linguistic criterion in the definition of confabulation, 18, 165, 198, 200–201
Lobotomy, 32, 77–78, 81, 98, 152, 184
Localizationism. *See* Holism vs. localizationism
Lying, 2, 15–16, 20, 37, 42, 72, 74, 92, 103, 112, 138
  and self-deception, 215, 220–221, 234
  and skin-conductance response, 93–94, 99, 170

Mamillary bodies, 8, *47(#13)*, 51–55, *53(#20)*, 56
Materialism, 39–40
Medial network, 82–83, *84*, 85, 87–89, 95, 127–128
Medial temporal lobe amnesia, 43, 49, 65, 191
Medial temporal lobe memory system, 44, 48–49, 88
Mediodorsal thalamic nucleus, 8, 35, *53(#7)*, *54*, 71, 87, 109, 146, 170, 228
  and memory confabulations, 51–55, 63, 69
  and orbitofrontal cortex, 77, 81–82, 85, 89, 181
Memory, 38, 43–49, 65–67, 193. *See also* Autobiographical memory; Semantic memory; Source memory; Procedural memory
Memory criterion in the definition of confabulation, 17, 198, 201–202
Mind reading, 42, 101–114, 148, 150–152, 155, 243
  and deception, 219–221
  and definition of confabulation, 177, 190, 192–193, 202, 210
  and self-deception, 212, 233–234
Mind-reading theory of the misidentification syndromes, 122–132

Mirror-image syndrome, 21, 97
  of anosognosia (hypochondria), 149
  of Anton's syndrome, 22
  of confabulation (OCD), 21–22, 97–99, 134, 215
  of prosopagnosia (Capgras' syndrome), 22, 118–119
  of self-deception (OCD), 234–235
  of sociopathy (OCD), 98
Mirror neurons, 106, 111, 113–114, 151
Misidentification syndromes, 114–117, 152, 155, 166, 177, 202, 210
Misoplegia, 136
Mnemonic concept of confabulation. *See* Confabulation
Moore's paradox, 188
Motor areas, 35, 73, 82, 87, 106, 111, 113

Neglect, 10, 18, 135, 140, 143, 150, 171, 224, 231
  and definition of confabulation, 180, 185, 195
Neurocomputation, 29

Obsessive-compulsive disorder (OCD), 22, 77, 97–99, 149, 175, 212, 215. *See also* Mirror-image syndromes
  and definition of confabulation, 183, 189, 194, 206–207
Occipital lobes, *33*, *34*, 49, 146, 182
Optic chiasm, 58
Orbitofrontal cortex, *33*, 35, 37, 42, *54*, 61, 63, 69, *75*, *83*, 101, *107*, 243
  anatomy and physiology of, 71–90, 95, 167
  and aneurysm of the anterior communicating artery, 57–58
  and anosognosia, 144–146, 151
  and consciousness, 174–175, 241
  and definition of confabulation, 179–182, 207
  integration of left and right, 155, 158, 168–169

Orbitofrontal cortex (cont.)
and Korsakoff's syndrome, 52, 54–55
and mind reading, 102, 104, 106–113, 121, 134
and the misidentification syndromes, 104, 116–117, 124–128
and obsessive-compulsive disorder, 97–99
and self-deception, 215, 228, 234
Orbital network, 82–83, *84*, 85, 88–89, 127–128
Out-of-body experience, 151

Paranoia, 22, 184, 232
Parietal cortex, 33, *34*, 36–37, 79, 110–111, 124, 139, 143–145, 163. *See also* Inferior parietal cortex
Perception-action cycles, 32, 73, 172
Phantom limbs, 139, 142
Philosophy, 37–40
Plausible confabulation. *See* Implausible vs. plausible confabulation
Polygraph, 74, 93–94
Posterior cingulate cortex, 87, 111, 144–145
Pragmatic errors, 105–106, 112, 203
Prefrontal cortex, 77, 82, 87, 89. *See also* Dorsolateral prefrontal cortex; Orbitofrontal cortex
Prefrontal lobotomy. *See* Lobotomy
Probability, reasoning about. *See* Estimation
Procedural memory, 46, 50
Process reliabilism. *See* Reliabilism
Provoked vs. spontaneous confabulation, 7–8, 17, 183, 191, 195–196
Prosopagnosia, 22, 118–119, *120*, 121–122, 130
Pseudopsychopathic syndrome. *See* Acquired sociopathy

Rationality, 188–189, 242–243
Reality monitoring, 45, 61–63, 91, 168, 175
Reduplicative paramnesia, 116, 118, 125, 130, 152, 162

and definition of confabulation, 177, 189, 191
Releasing effect, 32, 72, 184
Reliabilism, 207–208
Representations, 61, 150, 178, 185
Riddles, 40
Right hemisphere, 3, 11, 94, 112, 116, 118, 125, 150, 155–157, 164–170
and definition of confabulation, 180, 211
and anosognosia, 135–136, 140–143, 232

Sarcasm, 112
Schizophrenia, 8, 13, 15, 116, 122, 183, 186
Scientific method, 40–41, 239
Self, 5, 241–242
awareness, 61, 71, 78
monitoring, 61, 71, 169, 178, 185, 191
Self-deception, 41–42, 189, 212–218, 226–236
compared to confabulation, 225–228, 240
definitions of, 221–225
Semantic memory, 8, 45, 48, 64–65, 129, 177, 201
Shallow deception. *See* Deep vs. shallow deception
Simulation, 73, 103, 110, 113–114, 124, 126, 131, 133, 169, 233. *See also* Mind reading
Skin-conductance response (SCR), 22, 79, 92–97, 99, 102, 141, 144, 180, 229. *See also* Lying
and face perception, 118–119, 121
laterality of, 94–95, 155, 166, 168, 170
Social behavior, 51, 77–78, 120
inappropriate, 56, 102–104, 108, 112, 134, 145
and the orbitofrontal syndromes, 72, 74, 76, 80, 90, 99
Sociopathy, 42, 104, 133, 134, 212. *See also* Acquired sociopathy

and definition of confabulation, 180, 184, 190
as orbitofrontal syndrome, 71, 74, 78, 90–93, 99–102
and self-deception, 214–215, 228, 233
Somatosensory areas, 35, 59, 82, 106, 110–111, 139–140, 150, 190. *See also* Body representations
Source memory, 9, 48, 157, 167
Spatial disorientation. *See* Disorientation
Split-brain patients, 10–11, 14, 17, 42, 58, 142, 152–157, 162–164, 234
and definition of confabulation, 177, 193, 202
Spontaneous confabulation. *See* Provoked vs. spontaneous confabulation
Standards of quality, 81, 243
Striatum, 81, 87, 89–90. *See also* Caudate nucleus
Superior temporal sulcus, 35, 82, 88, *107, 111,* 168, 181
and anosognosia, 143–144, 151
and face perception, 120–121
and mind reading, 106, 110–111, 130, 152

Temporal lobes, *33, 54, 75,* 88, 97, 99, 109, 112, 151, 158, 160, 182, 211. *See also* Fusiform gyrus; Superior temporal sulcus
inferior temporal lobes, 82, 88, 121, 128, 136
and memory, 49, 55, 58
and misidentification syndromes, 116, 119, 124–126
temporal pole, 88, 109, 158
Tension, 74, 93, 214, 226, 230–231
Thalamus, *34, 36,* 52, 86, 139, 175. *See also* Anterior thalamic nuclei; Intralaminar thalamic nuclei; Mediodorsal thalamic nuclei
Topographic maps, 35, 181
Truth criterion in the definition of confabulation, 208

Two-phase theory of confabulation, 44–45, 61, 65–66, 146–147, 181, 190–192

Ventral stream, 33, 35–36, 81–82, 106, 110, 112–114, 118–119, 122, 151, 182
Verification processes. *See* Checking processes
Vivid day-dreaming, 59, 242

Wada testing, 165, 168
Weakness of the warrant, 213, 225
Weakness of the will, 232–233
Wernicke's aphasia, 202–203
Will, 170–174
Wisconsin Card Sorting Test, 45, 172
Working memory, 36, 49, 88, 98